George Herbert Little

The Marine Transport of Petroleum

book for the use of shipowners, shipbuilders, underwriters, merchants, captains,

and officers of petroleum-carrying vessels

George Herbert Little

The Marine Transport of Petroleum
book for the use of shipowners, shipbuilders, underwriters, merchants, captains, and officers of petroleum-carrying vessels

ISBN/EAN: 9783337092016

Printed in Europe, USA, Canada, Australia, Japan

Cover: Foto ©Andreas Hilbeck / pixelio.de

More available books at **www.hansebooks.com**

FIG. 66.

TANCYES BIRMINGHAM

From Tangyes Copyright Catalogue

THE

MARINE TRANSPORT

OF

PETROLEUM.

A BOOK FOR THE USE OF SHIPOWNERS, SHIPBUILDERS,
UNDERWRITERS, MERCHANTS, CAPTAINS,
AND OFFICERS OF

PETROLEUM-CARRYING VESSELS.

BY

GEORGE HERBERT LITTLE,

ROYAL NAVAL COLLEGE CERTIFICATE FOR
PRACTICAL PHYSICS, NAVAL ARCHITECTURE, MARINE SURVEYING, STEAM,
AND STEAM ENGINE;
BOARD OF TRADE CERTIFICATE AS MASTER MARINER;
EDITOR OF THE 'LIVERPOOL JOURNAL OF COMMERCE.'

FIAT LUX.

E. & F. N. SPON, 125, STRAND, LONDON.
NEW YORK: 12, CORTLANDT STREET.
1890.

PREFACE.

In submitting this work to the favourable consideration of those for whom it is intended, the writer disclaims at once any pretence to original research on his own part. Having been engaged in the petroleum trade, he found that among those chiefly interested there was not that amount of technical knowledge which the nature of the trade demands. Petroleum was regarded by some either as an exceedingly dangerous liquid which required most extraordinary precautions, or else it was treated by others in a reckless manner, which, as the reader is aware, has produced in some cases most lamentable results. It occurred to the writer that there was a want of some treatise which would be acceptable to the trade at large. He has endeavoured in an imperfect manner to supply this want. Knowing the regrettable distaste that masters and officers of ships have as a rule for anything which savours of what is called "mere theory," he has endeavoured to avoid this, and where he has found it impossible to do so he has, in deference to their feelings, been as lenient as possible. At the same time he would take leave to remark that, in this age of

technical education, we require an amount of knowledge, on the part of those occupying responsible
positions, which sadly interferes with the time-
honoured traditions of the sea.

The public demands from those to whose care is
entrusted the lives of their fellow citizens, at least a
fair knowledge of the properties of the cargoes they
carry. Every explosion of petroleum that has
hitherto taken place on board ship, has been due to
the neglect of those precautions which a slight know-
of petroleum would suggest to people of ordinary
intelligence. While on the one hand no amount of
reading will make a careless man careful, either of
his own or the lives of his fellows, yet on the other
there is, the writer is pleased to say, a growing
demand in the mercantile marine for the fullest
information about cargoes which under certain circumstances may be a source of danger or pecuniary
loss to those concerned. The writer's object has been
to supply, in an admittedly imperfect manner, some
information on the Marine Transport of Petroleum,
which will be useful to all engaged in the trade. With
the great development in the petroleum trade which
is certain to take place within the next few years,
there will be of necessity a corresponding increase
in the fleet of tank steamers. The writer views with
regret that the British shipowners have not as yet
shown very much disposition to follow up this important branch of their business. This is probably due
to the general but erroneous belief that a tank steamer

is useless for carrying any other cargo, and also to
the growing restrictions with which the authorities
surround a new business having in it any element
of risk to life and property. The writer has endeavoured to prove that handling petroleum is not
more dangerous than any other cargo, provided
always common sense and intelligence are used by
those in charge.

In putting together the following notes the author
has been much assisted by Sir E. J. Reed, F.R.S.,
M.P. ; Mr. W. H. White, R.N., Chief Constructor to
the Admiralty ; Professor Jenkins, M.A., of Glasgow
University ; Mr. B. Martell, Chief Surveyor of Lloyd's
—all of whom have permitted the author to make
excerpts from their well-known works. The author
takes this opportunity of expressing his thanks to the
foregoing authorities, to whom he is indebted for much
information. As regards the remarks on the nature,
composition, testing, &c., of petroleum, he is indebted
to Professor Tate, F.I.C., of Liverpool, who has taken
great interest in the subject, and is a well-known
authority on Chemical Technics ; and Mr. Boverton
Redwood, F.I.C., F.C.S., is another authority that the
author is deeply indebted to for much practical and
useful data.

Much useful information has also been derived from
those high-class journals *Engineering* and *Engineer*. In
fact the author has done little more than collect and
collate the data relating to petroleum which are found
in many publications, and arrange them for the use of

those engaged in the trade. Since Mr. Charles Marvin
has done so much to extend our knowledge of the
sources of petroleum in different parts of the world,
we may expect to see the industry assume the most
gigantic dimensions. Mr. Marvin may fairly claim to
have done more than anyone else in building up the
petroleum-in-bulk transport system.

The remarks on electric lighting are the results of
the author's practical experience as an electrical engineer while in the service of a distinguished firm of
electric lighting engineers. In conclusion, the author
trusts that, notwithstanding the many blemishes that
the work contains, it may prove of some slight service
to those for whom it is intended.

GEO. HERBERT LITTLE.

7, VICTORIA STREET,
    LIVERPOOL.

CONTENTS.

—◦◦◦—

CHAPTER I.

CHAPTER II.

CHAPTER III.

CHAPTER IV.

PUMPS—LOADING—DISCHARGING—BALLAST-TANKS—
VENTILATION.

CHAPTER V.

TESTING.

CHAPTER VI.

THE LIGHTING OF PETROLEUM STEAMERS BY ELECTRICITY.

CHAPTER VII.

USEFUL TABLES, DATA, ETC.

THE

MARINE TRANSPORT

OF

PETROLEUM.

———•———

CHAPTER I.

1. PETROLEUM, literally "rock-oil," is the name
given to a liquid which is found in most countries of
the world. As its name implies, it is of geological
origin, but geologists are not agreed as to the primary
causes of its formation and distribution. Professor
Mendeleef has investigated the question, and has succeeded in producing petroleum artificially by acting
on incandescent metallic carburets, especially that of

B

iron, with water, which is decomposed in the process, the oxygen uniting with the metal forming an oxide whilst the hydrogen combines with carbon, forming liquid and gaseous hydrocarbons which cannot be distinguished from the natural petroleums. He consequently attributes the production of the latter to water trickling through fissures in the earth's crust, till it reaches a stratum of carburet of iron, when it is decomposed as described above. Owing to the high temperature, even the heavy hydrocarbons will be given off in a state of vapour, and will therefore rise to a higher level, where the less volatile will be condensed to liquids, the whole being stored up till released by the explorer's drill. Should any considerable portion of the lower strata of the earth's crust consist of metallic carburets, Professor Mendeleef is of the opinion that the action may be continued almost indefinitely.

2. Petroleum exists either as a gas, a liquid, or a solid. In one or other of these forms it is found in many parts of the United States, Canada, South America, South-eastern Europe, South Russia, Burmah, Egypt, Australia, and New Zealand.

As a gas, petroleum (in this form called natural gas) is found in the United States, where it exists stored up in the bowels of the earth, in what might be called natural gasometers. It is apparently generated from crude petroleum, and as it accumulates it exerts an immense pressure upon the surfaces of these natural gas-holders.

When, as in the operation of boring, the reservoir is pierced, the confined gas rushes up with such violence as often to blow out the boring plant. The pressure of the gas, as it issues, has been measured, and at one well in the United States a pressure of gas was registered of 187 lbs. per square inch.

This natural gas is used for purposes of illumination and heating. Many towns in the oil region of the United States almost entirely depend upon the natural gas for fuel and light.

In its solid form, petroleum exists as a wax known as "ozokerit,"* extensive deposits of which are found in Gallicia, and in the island of Tcheleken in the Caspian Sea. This wax is largely used in the manufacture of candles.

The firm of Messrs. Price & Co., of London, use immense quantities in the manufacture of their well-known "ozokerit candles."

3. As a semi-solid, petroleum exists as bitumen, lakes of which are found in South America and California. It is also found in Syria and Persia. The Biblical mention of pitch undoubtedly refers to bitumen. The crude product which floats upon some of the marshes in Persia, Syria, and Irak-Arabi is nothing but petroleum ; it has a very offensive smell, and hence the personal defilement of those who touched "pitch" is easily explained.

* Ozokerit, or ozokerite, or "earth wax," consists of hydrogen 13·75, carbon 86·25. A most useful substance, much used, in conjunction with gutta-percha and indiarubber, as an insulator. It is similar to "paraffin" wax in some respects.

In the construction of the Ark, bitumen is most probably meant as the material wherewith to "pitch it within and without."

We may soon expect to see this form of petroleum assume a value that at present it does not possess. In many parts of Great Britain, extensive deposits of what is known as "bituminous shale" exist. This substance may be regarded as a form of solid petroleum, and is used in the manufacture of shale oil. The old-established and well-known firm of "Young's Paraffin Oil Company" manufacture their paraffin exclusively, we believe, from the shale deposits.

4. Petroleum in its liquid form, as it issues from the wells, is a very complex combination of various chemical compounds. It is a member of that large class of compounds called hydrocarbons. Petroleum roughly consists of about 85 per cent. of carbon and 15 per cent. of hydrogen. Impurities, in the form of nitrogen, oxygen, and sulphur, are also generally present. It may be well to describe the two elements carbon and hydrogen.

5. Carbon (C) is the name given to pure graphite or plumbago, pure charcoal, lampblack, and the diamond. It is the principal constituent of coal. Combined with the gas oxygen in the proportion of two atoms of oxygen to one of carbon it produces carbonic dioxide (CO_2). The latter is the gas exhaled during respiration, and is poisonous. Carbon combined with oxygen in equal proportions forms carbonic oxide (CO), one of the gases given off

when charcoal or coal are consumed in a furnace. Carbon requires a plentiful supply of oxygen for its combustion, but during the process of combustion carbonic dioxide and carbonic oxide are formed. The former gas, carbonic dioxide (CO_2) is a most excellent means for preventing combustion. Carbon combines with hydrogen gas in various proportions to form the numerous bodies called hydrocarbons. Petroleum, naphtha, sugar, coal, alcohol, ether, chloroform, morphia, strychnia, are all examples of the combination of hydrogen with carbon.

6. Hydrogen (H) is the lightest body in nature. It is very combustible, but does not support combustion ; that is, it requires the presence of another body (oxygen), before it will burn.   It is an invisible gas, like oxygen.

Hydrogen combined with oxygen in the proportion of two atoms of the former to one of the latter forms water (H_2O) ; combined with carbon it forms, as before stated, the numerous order of hydrocarbons. Thus we have a very light invisible gas, hydrogen, combined with the solid carbon, forming the liquid petroleum.

7. Petroleum as it issues from the wells varies greatly in its chemical composition, colour, odour, and specific gravity, the last varying from ·780 to ·970. From the crude oil the gases methane and propane are given off. Methane (CH_4) is that gas, known as " marsh gas," which is given off from stagnant pools in warm weather, and when burning, as it

does, with a pale blue flame is known in the country
as the "Will-o'-the-wisp." It is also present in coal-
mines, and when issuing from a seam it mixes with
the air, forming that dangerously explosive mixture
called fire-damp. Methane is one of the constituents
of the common coal gas used for illuminating
purposes. Ethane (C_2H_6) and propane (C_3H_8) are
important from a chemical point of view, but need not
be further discussed here. Butane (C_4H_{10}) is another
gas found in cannel coal and petroleum.

8. Petroleum is the generic name given to a large
family of hydrocarbons, all of which have certain
characteristics in common but differ very much from
each other in colour, specific gravity, inflammability,
odour, and quality. Thus we have from crude petroleum the olifine series of compounds, the paraffin series,
the benzine series, and other groups of compounds
all of which find a use in the hands of the chemist.

9. From the wells the crude petroleum is pumped
into storage tanks, and from thence into retorts in
which it is distilled, the object being to separate
the various kinds of oil most suitable for lighting,
heating, and lubricating purposes. Without going into
details, the process of refining and distillation is somewhat as follows :—The crude petroleum is led into a
retort and heat is applied ; the volatile products known
as the naphtha and benzine series are given off as gases,
and are condensed in another chamber. The heat is
not allowed in this method to exceed a certain temperature. After the more volatile products are given

off the oil in the retort is called " distillate "; it is then either shipped to a refinery, as is the case with much of the distillate from Baku, or it is refined on the spot. To refine it, it is mixed and agitated with hydric sulphate (sulphuric acid), which removes the olifine and other non-saturated hydrocarbons, and then with hydric nitrate (nitric acid). The hydrocarbons of the benzine series, and all other compounds except the paraffins, are thereby oxidised and converted into nitrocarbons. The oil is then washed with water and caustic soda, and after standing on settling tanks it is ready for shipment as refined kerosine.

10. According to Mr. Redwood, Russian crude oil yields about 30 per cent. of kerosine. According to an American authority, 100 kilos of Russian crude petroleum give 45 per cent. of refined oil; 100 kilos of Galician crude petroleum give 50 to 52 per cent. of refined oil; and 100 kilos of American crude petroleum give 60 to 63 per cent. of refined oil.

11. In the Russian oil region the process of continuous distillation is adopted. A number of stills or retorts, generally about 25, are set up, connected with each other, and a continuous stream of oil flows through the whole. The crude oil entering the first gives up the more volatile constituents, and passing into the next retort gives up the less volatile constituents, and so on, till in the last retort it becomes the residuum or " astaki," also called " masut," specific gravity about ·903.

In the American method of refining oil, the crude

oil is very slowly distilled, and instead of allowing the volatile portions to pass over and condense in
another retort they fall back into the heated oil, and by
being subjected to a heat above their boiling-point they
become broken up into other compounds. This process
is called " cracking," or " destructive distillation," and it
enables a larger yield of kerosine to be obtained—as
much as 70 per cent. This, however, is not indicative
of the value of American oil as compared with Russian.

12. Crude petroleum from the United States,
according to the rules of the New York Produce
Exchange, is understood to be pure natural oil,
neither steamed nor treated, free from water, sediment,
or any adulteration, and of a specific gravity of 43°–
46° Beaumé ($\cdot816°-\cdot794°$). Naphtha is required to
be " water-white," and sweet, and of specific gravity
from 68°–73° Beaumé ; while the residuum is understood to be the refuse from the distillation of crude
petroleum, free from coke and water, and from any
foreign impurities, and of specific gravity from 16°–
21° Beaumé ($\cdot96-\cdot93$ specific gravity). Owing to the
present high state of perfection attained in refining
and distillation, the well-known Standard Oil
Company produce high grade oils for lighting
purposes that cannot be surpassed either for colour,
smell, high flashing and firing point, and low specific
gravity. This last is a very important point, since
the supply of oil to a wick depends upon the specific
gravity of the oil and the capillarity of the wick.
With two wicks precisely similar, but using oils of

different specific gravity, that wick which is using the lighter oil will, other things being equal, give better light than that which burns the heavy oil. It is the comparatively low specific gravity of the Russian oils that has prevented them being used in lamps, but lamps are now constructed for the purpose of burning the denser oils.

13. Paraffin is often confused with petroleum : it really is one of the products of petroleum. It derives its name from *parum affinis* or chemical indifference. In the process of distillation of crude petroleum, the residuum, after the removal of the illuminating oil and naphtha, is distilled, and a product or rather a group of hydrocarbons is obtained called the paraffin series. After the removal of the paraffin the result is paraffin oil. The crude paraffin oil is placed in freezing houses, and after being subjected to a very low temperature and allowed to remain at rest, solid paraffin crystallises out. This last product is an extremely useful substance. It forms one of the best insulating bodies known, and is largely used in the construction of electrical instruments. As a lighting agent it is used for candles.

14. As regards the safety of commercial petroleum, Messrs. Newbury and Culter say, that the low flashing point and danger of oils is not necessarily dependent upon the presence in them of small quantities of very volatile substances, but may also be caused by a large proportion of soluble high-boiling constituents. The method by which the igniting

point of an oil exposed in an open vessel is determined, is generally held to afford no accurate information as to the safety of the oil in question.   In
spite of this, however, the sale of any oil which
ignites below 300° F. (149° C.) is forbidden in the
United States.   An oil may show a very high igniting
point and yet possess a low flashing point, and consequently be exceedingly dangerous.   This is especially the case with oils which consist of a mixture
of very heavy and very light hydrocarbons.   The
more uniform is the composition of an oil, the nearer
do its flashing and igniting points approach.

The ignition point of an oil depends, not only on
the proportion in which the light constituents are
present, but also on the properties of the oil as a
whole.   A dangerous oil may be made to satisfy the
legal requirements by the addition of a sufficient
quantity of lubricating oil, only such a mixture of
very different substances possesses very small illuminating power.   It is doubtless this circumstance which
has moved the manufacturers to the oft-repeated
complaint that a high flashing point is not attainable
in conjunction with good burning qualities.   If the
refiners would content themselves with obtaining
rather less oil, by rejecting a portion of the benzine
and of the lubricants, they would have no difficulty
in producing illuminating oils of high flashing point
and excellent quality.   The well-known "Astral oil,"
which flashes at 125° F. (52° C.), is a standing protest
against this claim of good burning power for oils of

low flashing point alone. It is simply, as it were, the kernel of the burning oils, freed both from light and heavy constituents. Experiments instituted by these gentlemen have shown that the oil in question burns excellently in ordinary lamps, and its photometric examination has given results which place it on perfect equality with any commercial oil.

The refiners can scarcely be expected to produce oils of higher flashing point than is required by law. The raising of the legal limit to 120° F. (49° C.) would not, however, be inflicting a hardship on the manufacturer, and would insure a safety of life and property which would far outweigh the slight increase in the cost of illuminating oils. It has been thought that the effect of such a law would simply be to make the manufacturers add heavier constituents to their oils, and thus satisfy all legal requirements at the expense of the burning qualities of the oil. This would, however, be in direct opposition to their own interests, since oils of high flashing point and excellent qualities are already largely sold and burned.

Finally, any oil is dangerous when heated above its flashing point, be that point high or low. There can, therefore, be no doubt that when ordinary petroleum is employed in the very powerful lamps now in vogue, our safety depends to a great extent on such an arrangement of the parts of the burner as will render the ignition of explosive vapour within the lamp as difficult as possible. An oil which flashes below 120° F. (49° C.) cannot be burned in such lamps with

safety, and when oils of ordinary quality are employed, the conditions necessary for an explosion are
always present.

15. The great commercial value of petroleum lies in
the fact of its being an abundant and cheap source of
light and heat.   Modern Physical Science teaches us
that light, heat, and work are really synonymous : that
is, to do work of any sort, whether bodily, mental, or
through the medium of an engine, we must have the
expenditure of heat in some way—that is, we must
have combustion of some material.   Of course, combustion is not necessarily attended with the presence
of flame.   We have combustion in different forms
going on in our bodies ; in fields ; in electric batteries ;
in the forest, where dead leaves are being resolved
into their elements.   For the present purpose, combustion may be described as the process whereby the
energy contained in a body is made to give it out in
the form of heat, and finally as work.

It is easily understood that different kinds of fuel
vary greatly in the amount of work they are capable
of doing.   The value of a fuel is measured by the
number of pounds of water which can be evaporated
per pound of fuel consumed.   1 lb. of the best Welsh
coal, when burnt with perfect combustion, is capable
of evaporating 9·5 lbs. of water.   1 lb. of the refuse
of petroleum, called " astaki," is capable of evaporating
from 12 to 14·5 lbs. of water according to the method
of combustion.

In the combustion of coal in a furnace there are

serious losses of heat : every time the door is opened cold air rushes in, and every shovelful of fuel put on for a time damps the flame and hinders combustion. Whereas by burning petroleum in a suitable furnace, a continuous feed can be maintained, and a body of flame will continuously impinge upon the furnace, and every scrap of useful heating surface can be used, which is not the case with coal.

16. Few by-products are of such use as astaki ; and it is strange that, in this age of economy, shipowners, who are continually complaining of foreign competition, foreign bounties, loss of bargaining power, and so forth, should allow German and Russian engineers to show us the road in the matter of liquid fuel.

Astaki is used as fuel in the Russian railway engines and also in the Caspian steamers ; a portion of it is converted into lubricating oil, and another portion into an oil similar to mineral sperm obtained from American petroleum. Astaki yields about 30 per cent. of lubricating oil, and about 12 per cent. of mineral sperm or colza. This latter kind of oil is called also " pyronaphtha," spr. g. 865, and fire test 265° F.; flashing point, 230° F. open test, or 205° F. closed test.

From astaki are likewise obtained tar, naphthaline, "green grease," and pitch, and an illuminating gas of very high order. Thus it will be seen what an immense field of industry is opened up by the gradual development of the enormous natural storage tanks that exist in almost every part of the world. Thanks

to the researches of Mr. Chas. Marvin we can con-
template the exhaustion of our coal beds with perfect
equanimity.

Liquid fuel is slowly but surely making its way as
a substitute for coal in steam vessels ; one or two
vessels trading regularly to Batoum have had furnaces
fitted for its combustion, but hitherto these efforts have
not been crowned with success in British vessels.   In
the writer's opinion this was largely due to the fact
that the engineers employed in testing its practical
value on board ship were in most cases sadly deficient
in the necessary technical knowledge.   In fact the
carriage and combustion of petroleum is a thing that
requires " living up to," just like forced draught.   It
has to overcome a vast amount of ignorance and
prejudice at the hands of those who should be the
most eager to take advantage of an improvement.

In introducing any new improvement on board
ship, it is too often forgotten that success or failure
really is a question that is decided by those who have
to manipulate it.   In the case of liquid fuel it was only
to be expected that the usual run of engineers would
pronounce against it, as there was an element of
uncertainty about petroleum which caused it, and in
fact does cause it, to be regarded by partially educated
people with suspicion.   In those cases where steps are
taken to ensure its handling by technically-educated
people the burning of astaki is attended with the
happiest results.

In the hands of Mr. Urquhart, C.E., the problem of

burning petroleum refuse (astaki) in locomotives has been completely solved with the most satisfactory results.

17. The progress of liquid fuel is being attended with the greatest success in America; and as our information on the point is derived from the pages of that high-class paper *Engineering*, we cannot do better than give the following extract from that journal :—

"From the Lima oilfields, Ohio, to Chicago, is a distance of 270 miles, and last year the two were successfully connected by an 8-in. pipe, worked by a mammoth pump. In the meantime the Standard Oil Company had prepared the public mind at Chicago for the innovation by establishing a Liquid Fuel branch of its business, and organising a staff of engineers and workmen capable of fitting oil-fuel appliances copied from those in vogue in Russia, but in many instances improved upon; and the result was, as soon as Lima oil began to flow into the reservoirs at Chicago, numerous applications were received from local factories to fit up oil appliances and furnish a supply. Since then the movement has rapidly extended, and ramifications of oil pipes are driving out coal from every quarter of the city. As might be expected, the coal trade in consequence has been "very depressed," and is not likely to undergo an improvement, seeing that public feeling is in favour of the change owing to the greater purity of the atmosphere arising from the use of the new fuel. A

short time ago Pittsburg, formerly one of the dingiest
cities in the world, became transformed into a clean
one with a pure atmosphere, by the adoption of
natural gas instead of coal.   If liquid fuel becomes
generally adopted at Chicago, an improvement just
as great will be experienced there, and the innovation
will doubtless spread in time to other American
cities ; Russian competition having no effect on
this   branch   of   the   Standard   Oil   Company's
business.

" This use of petroleum for metal-working is attended
by circumstances which render it of so great value in
comparison with coal, as to be almost independent of
the cost of fuel, because the flame is a deoxidising
one, and therefore there is no liability of burning the
steel.   Also   the   proportion   of   imperfect   work, or
" second," in cutlery establishments where this oil fuel
has been introduced, is reduced to almost nothing.
In addition to the deoxidising flame, the heat can be
controlled at the will of the operator to suit the
especial things in hand, and therefore expedites work
to a great degree.

" The conditions of supply and use, as well as the
mechanical properties of petroleum in the United
States, are so different from those in Russia as not to
afford any sound basis for comparison.   Suffice it
to say that the present consumption of fuel oil is
believed to be less than half of the supply from the
oil district referred to, and that the price is now in
excess of coal, except in a few favoured localities

where the position is such that the price of coal fuel is relatively high and oil fuel relatively low.

"There are three methods recognised in general use for the combustion of petroleum. One consists in its reduction to gas by destructive distillation in a gas plant, after which it is burned under boilers in a manner comparable to natural gas. Another method is by forcing the oil under a boiler in a spray by means of compressed air ; and a third, which is the more general in its application, uses an injector which is operated by a jet of steam from the boiler to throw the oil into a furnace, where it is vaporised and mingles with air, which is also thrown from the injector.

"The general introduction of petroleum fuel also involves questions relative to the storage of petroleum in cities and around manufactories. All authorities agree that the hazard attending the use of petroleum for fuel is largely of a controllable nature, being dependent upon the precautions taken in regard to its storage and use. The method most highly recommended consists in placing the main tank of the system at some place where it can be filled by gravity from the tank-cars used for the transportation of crude petroleum. It is advised that storage tanks should be placed in the ground, in order to avoid extremes of temperature and hazard to property by the boiling over of the oil in case of fire and they should also be provided with a ventilating tube to carry off hydrocarbon vapours arising from the oil.

C

From these main tanks a supply is taken to a feeding
tank by means of a pump; the second tank being
placed at such a height that its contents can flow
back therefrom directly, the object being to conduct
any excess of oil at the close of the day back to the
main storage tank.

"Whenever the oil is ignited under a boiler it is
necessary to apply the flame in the furnace first, and
then turn the oil on to it ; because if the oil is turned
on first, and the fire applied, it is more than possible
that some portion of the air in that furnace will be
commingled with vapour of the oil in just the proportion necessary for an explosion ; or if it does not
conform exactly to these proportions, it will, in almost
every instance, form a mixture near enough to an
explosive one to produce a very sharp puff.   The
various connections between the tanks should be
provided with double valves, so that if the oil cannot
be shut off in one place it will be possible to do so at
another.

"Combustion of petroleum, like that of all gaseous
fuels, can be carried on with a much higher degree
of efficiency than that of solid fuel ; while the ratio
of calorific value of a pound of oil is $1 \cdot 37$ to that of a
pound of bituminous coal, and $1 \cdot 61$ to that of a pound
of anthracite coal, yet in actual practice it is found
that 1 lb. of oil is fully equivalent to $1\frac{8}{10}$ lb. of coal.

"There are three forms of advantage in the use of
oil, apart from this direct cost of fuel.   These relate
to the economy of labour in attendance at the boiler :

to the larger consuming capacity of a boiler, because the heat is applied continuously without any interference or interruption comparable to that of applying the coal on a furnace of an ordinary boiler ; and there is an absence of smoke and of cinder, or any residual product connected with a fire. In the case of its application to locomotives, it is alleged that the freedom from cinders has diminished the wear upon the valve gear."

Looking at the great evaporative efficiency of petroleum as compared with coal, and bearing in mind that while the latter occupies from 40 to 42 cubic feet the former occupies but 39, it will be seen that from the shipowner's point of view, astaki is the ideal fuel. The problem as to the best means of combustion is solved, and it only requires the expenditure of a little enterprise on the part of shipowners, and the erection of astaki storage tanks at the coaling stations, to render its use possible on ordinary cargo and passenger vessels.

18. The public have been from time to time startled by the explosions that have occurred on board the *Petriana* and *Ville de Calais*, and on board small vessels which were carrying petroleum as cargo in barrels. With regard to all these cases, ignorance and carelessness were, as usual, the causes of much loss of life and property. The danger of petroleum vapour is just the same as that attending spirit vapour, and a ship laden with petroleum in barrels is subject to the same risk as attends the

C 2

storage of hundreds of puncheons of whisky or rum in the bonded warehouses. In the case of the coaster *United*, which blew up at Bristol, the cargo consisted of the most volatile product of petroleum stored in wooden casks!

The *Engineer* makes the following remarks on this case, which we reproduce :—

"The vessel concerned in the Bristol explosion was of small size, having a registered tonnage of 58 tons, or 64 tons gross. At the time of the explosion she had her cargo on board, consisting of 310 barrels of light mineral oil of the kind which Colonel Majendie designates 'mineral spirit,' being in fact a species of benzoline, known in the trade as Pratt's deodorised naphtha. The voyage was to be from Bristol to the river Thames, and it might easily have happened for the explosion to occur at the end of the trip instead of the beginning, in which case London would have been more intimately acquainted with the disaster. Such were the circumstances of the case, that the blowing up of this unlucky ship seems to have been tolerably well assured to occur at some time or other. The vessel was in no way adapted for her cargo, and the master had only a very hazy conception of the danger by which he was beset. He was overwhelmed with good advice and solemn admonitions, rather to his surprise and annoyance. 'He had been accustomed to carry dynamite, and never had so much fuss made about that.' The peril arising from the pervasive vapour of petroleum was a point which he never seems to have understood ; but the error is one which very generally prevails, and many lives have been lost in consequence. The flame-carrying power of petroleum vapour when mixed with air is extraordinary ; and the readiness with which the lighter kinds of petroleum give off their vapour needs to be more generally understood. At Colonel Majendie's request, Dr. Dupré carried out some experiments, by which it was found that naphtha like that on board the *United* was of such a nature that one volume of the liquid would render 16,000 volumes of air inflammable, and 5000 volumes strongly explosive. One volume of the liquid gave 141 volumes of vapour at an ordinary temperature, so that one volume of vapour would render inflammable about 113 volumes of air, and would give a strongly explosive character to about 35 volumes. It is mentioned as a circumstance admitting of no dispute, that benzoline in barrels is subject to a considerable and continuous evaporation. Colonel Majendie states that on more than one occasion, in his visits to petroleum stores, he has taken samples of the atmosphere, and in several instances has found these to be either explosive or inflammable. In like manner he took three atmospheric samples from the stores where a considerable quantity of naphtha was kept, of the same quality as that which had been delivered to the *United.* The barrels appeared to be in good condition, and the storehouse was a cool cavern, with a ventilating shaft. One of the three samples proved to be almost explosive, and the two others quite so. That there was a very decided escape of petroleum vapour into the cabin of the *United* is proved by known facts preceding the explosion. The entire cargo had been on board nineteen hours when the explosion occurred, thus affording time for the accumulation of the vapour. The capacity of the vessel when empty amounted to about 6000 cubic feet, and of this space something more than half was occupied with the cargo. The only ventilation was that given by the hatchways ; and as the specific gravity of the vapour was fully three-and-a-half times that of air, the amount of ventilation from this source was practically very small. It was owing to the boisterous state of the weather that the vessel was detained in harbour long enough for the explosion to take place before her departure. Of the extraordinary violence of the explosion, and the fierceness of the fire which followed, sufficient is already known, and the loss of three lives has to be deplored."

The disaster which befell the *United* was on a lesser scale than the Calais explosion. The *Ville de Calais* was a steamship of 1200 tons register, specially fitted with tanks and tubes for carrying the oil. She had discharged her cargo, consisting of crude petroleum, and water was being pumped into one of the tanks to serve as ballast, when by some means the explosive mixture of petroleum vapour and air which remained in the tanks was fired, and a tremendous explosion was the result. It is remarked by Colonel Majendie that the risk from fire and explosion is not limited to cases in which whole or considerable cargoes of petroleum spirit are shipped. A few barrels, or even one, may suffice. One gallon of petroleum spirit, it has been shown, is enough to render 16,000 gallons of air inflammable, representing a space exceeding 2000 cubic feet. The penetrating nature of the vapour increases the risk, a fact which has been proved by direct experiment, as well as indicated by actual misfortune. This quality, combined with the high specific gravity and flame-carrying power of the vapour when combined with air, renders its presence highly dangerous, even when the quantity may be small.

Another disaster attended with loss of life was that of the *Catherine*, a small wooden schooner of 85 tons. She had loaded a cargo of 367 barrels of petroleum spirit or benzoline, and was lying at anchor in the Thames preparatory to sailing. It being practically an impossibility to prevent evaporation from such a highly volatile liquid, the result was that vapour was disengaged, and mixing with the air in the hold, eventually formed a highly explosive gas. Notwithstanding that the captain had been warned not to permit any light or fire on board, the galley fire was lit, and the vessel blown up.

Another case under similar circumstances, was the loss of the sailing barge *Charles Little*, also laden with benzoline. It would be thought that the experience gained in the two former cases would

have been sufficient to give publicity to the precautions necessary to take in carrying such cargoes,
but it seems that little short of legislative measures is
effective to enable many people to grasp a few facts.

The lessons to be learned from these three cases
are :—

First, that the more volatile products of petroleum
should in no case be carried in wooden casks.

Second, small wooden vessels are unsuitable.

Third, that owners and charterers should take care
that those who are in charge of petroleum-carrying
vessels, should understand something about the
cargoes they carry.

19. In England we, with fine old crusted conserva-
tism, still cling to the time-honoured and familiar
wooden barrel, simply because we are used to it.
Such a method of transportation is neither economical nor efficient—in fact, for railway transportation,
it is positively dangerous. A fire, attended with loss
of life, took place some years ago through kerosine
being carried in barrels on a train which caught fire.
Had the oil been in properly constructed tanks, there
would have been not the slightest danger. It is
worthy of notice that one seldom or never sees a
barrel of oil which is perfectly dry on the outside,
there is always a slight amount of moisture, especially
near the chimes of the cask. Ordinary burning oil
can be carried in casks on board ships under proper
conditions with perfect safety, but the captain should
always satisfy himself as to the grade of the oil.

For the more volatile products of petroleum, casks are simply invitations to disaster, although for ordinary burning oil they do well enough; but for

FIG. 1.

purposes of distribution from storage tanks to the consumer, Phillips's tank-cart (Fig. 1) is a most useful and valuable improvement, and the use of which might, the author thinks, be rendered compulsory for the local transport of dangerous oils. The accompanying

engraving shows the features of this conveyance. The
writer cannot too strongly urge upon all concerned,
that under no circumstances should petroleum spirit,
benzoline, naphtha, &c., be carried in wooden casks.
The explosions at Bristol and at Thames Haven have
proved in the most practical way that casks for the
conveyance of the volatile products of petroleum are
as fitted for the purpose as a sponge is to transport
water. As a matter of fact, a wooden cask containing
petroleum spirit is every bit as dangerous as a sponge
saturated with spirit. The writer has many times
warned in the press those who, in despite of all
warning, will continue to expose the lives of their
fellows to unnecessary risk. Captains and officers
should refuse to take on board any oil unless they are
satisfied that it is safe.

20. Like all real improvements, the transport of oil
in bulk met with considerable opposition for some
time. " Old salts " said the ships were dangerous,
and by a peculiar process of argument the conclusion
was arrived at, among other dangers, that bulk
steamers were especially liable to be destroyed by
lightning.

An oil-tank steamship might possibly be struck
by lightning, but no instance has so far occurred,
although in storage tanks ashore cases are by no
means uncommon. An iron vessel in general enjoys
almost a perfect immunity from lightning discharges.
Sir William Thomson, perhaps the greatest living
authority on all subjects connected with magnetism

and electricity, says, that a sheet-iron house is the very safest place in a thunderstorm, and advocates that all powder magazines should be constructed of iron throughout. Considering that an oil-tank steamship is wholly constructed of iron or steel, and that it floats on an excellent conducting medium, we may safely dismiss all fear of explosion of oil or vapour through the agency of lightning.

Having regard then to the vast deposits of petroleum in various parts of the world, and the great value of petroleum as a source of light and heat, the time is not far distant when instead of coal depots, we shall have oil depots; and the problem of the next generation will not be as now the defence of our coaling stations, but the defence of our oil depots.   It is therefore, the writer thinks, very necessary that the question of economical and safe transport, should be thoroughly understood by owners, captains, and officers of petroleum-carrying steamers.

21. As regards the transport of the crude and refined oil from the sources of supply to the loading port, little need be said, as it does not fall within the province of this work ; it may, however, be dealt with in a brief manner. From a paper by Mr. J. Harris, of New York, read at the Aberdeen meeting of the British Association, we give the following excerpt :—

"The pipe-lines of America, with but a single exception, are all amalgamated in one Company, known as the National Transit Company. The length of the New York line is about 300 miles. There are two lap-welded wrought-iron tubes, 6 inches in diameter, the entire distance, while a portion of the way a third 6-inch pipe is laid. The number of pumping stations is eleven, they are about 28 miles apart, and are equipped with duplicate compound condensing pressure pumping engines, and the greatest elevation between stations above tide-water is 2490 feet. The Philadelphia line is 280 miles in length, diameter of pipe 6 inches, number of stations six. The Baltimore line covers 70 miles of direct pumping through a 5-inch pipe. The Cleveland line is 100 miles long, diameter of pipe 5 inches, with four pump stations. The Buffalo line is 70 miles long, diameter of pipe 4 inches, and it has two stations. The length of the Pittsburgh line is 60 miles, diameter of pipe 4 inches, with two pump stations. The entire length of these lines, including the duplicate pipes, is upwards of 1300 miles, while the length of the collecting lines of 2 inches diameter in the oil region, is estimated to be from 8000 to 10,000 miles."

The Worthington type of pumping engine is used exclusively on the pipe lines of the National Transit Company. The average amount pumped per day through the New York lines, is about 28,000 barrels, and the average head due to friction and elevation is about 900 lbs., sometimes rising to 1200 and 1500 lbs., depending on the piston-speed of the pump.

The engines on the main 6-inch lines, are from 600 to 800 horse-power, while those on the 4-inch and 5-inch lines are from 150 to 200 horse-power. The pipe is made especially for the service, is lap-welded wrought iron, and is known as oil-line pipe.   The lengths of 18 feet are fitted at each end with coarse and sharp-cut taper threads, nine to the inch, and with long sleeve couplings also screwed taper.   The taper is usually $\frac{3}{4}$ inch to the foot for 4-inch pipe.   The lines are laid two or three feet below the surface of the ground, and from time to time bevels are made in the pipe to allow for expansion and contraction.

At the different pumping stations are located one or more receiving tanks made of light boiler plate, dimensions 90 feet diameter by 30 feet high, and the oil is pumped from tanks at one station to tanks at the next, though there have been cases where loops have been laid around stations and oil has been forced a distance of 110 miles with one pumping engine.   Duplicate pumping engines are located in each station, so that there may be no cessation in pumping, one engine being in constant service.

A similar method of transporting the oil is proposed, and we believe is partly in operation, in South Russia. In South-eastern Europe petroleum is transported by rail in specially constructed cars—cylindrical tanks, 26 feet by 66 inches, made of light boiler plate, and mounted on eight-wheeled platform cars, and holding about 2000 gallons.

In South Russia, Mr. Tweddle says, the crude oil when pumped from the wells is run into small receivers known as "measuring tanks;" from these it is pumped by small branch lines to the central pump station, where it is stored in large iron tanks, the average capacity of which is 10,000 barrels (1500 tons). From these tanks it is drawn by the main pump and forced through the pipe-line towards its destination at the Tchornay Gorod refineries. This system of pipe-line transportation was introduced in the American petroleum fields some twenty-five years ago, and although great difficulties had to be overcome, the lines were increased in number and length until the present day, when there are over 10,000 miles of pipe used for this system of transportation. The tubes of which these lines are constructed are lap-welded, and vary from 2 inches to 6 inches in diameter, and are proved to a pressure of from 1500 lb. to 1800 lb. per square inch; they are furnished with a long taper socket, so as to have a perfect contact throughout the whole length of the thread, which for pipes over 2 inches in diameter number generally eight to an inch.

Worthington pumps are almost wholly used for forcing the oil through the pipes. What is required for the work is, that the pumps should work regularly at a pressure of from 1000 lb. to 1500 lb. per square inch, and few makers indeed are able to build a pump to meet this requirement; and in machines of inferior construction the "hammering" of the valves of the pump barrel can be heard for miles along the line. This constant jarring is very destructive to the whole line by its crystallising action on the iron, and is also liable to cause leaks at the joints. The general working pressure on a good line varies from 800 lb. to 1500 lb. per square inch.

22. On reaching the loading port the oil is pumped into storage tanks, which are similar in all respects to those at the discharging port. It is advisable that these tanks should not be erected near inhabited districts, nor anywhere in the vicinity of warehouses. In the writer's opinion they should be sunk some distance in the ground, in fact, built in a concrete basin. They should not be erected on an elevated position, and special precaution, should be taken against lightning. Pipes leading into the top of the tank for filling or ventilating purposes should, as regards the former purpose, be led down the side of the tank. Ventilating pipes and gas-escape pipes, should not project into the interior of the tank, because, supposing the tank to be three parts full, the space between the surface of the oil and the crown of the tank will be occupied by inflammable gas. If the ventilating pipe projects into the tank, and the former receives an electrical discharge through it, the current might pass through the crown and sides of the tank and make earth outside ; it *might*, also, jump across the space between the lips of the pipe and the oil, producing a spark of great length, and so explode the gas. The writer does not say that this would always occur, but when the curious vagaries of the lightning discharge are considered, and remembering also the enormous electromotive force of the discharge, such an occurrence is not improbable. As a matter of

fact, storage tanks have often been destroyed through the lightning discharge "arcing" across the space occupied by the gas. It may be here remarked that, in the event of a storage tank becoming the scene of a conflagration, water is not of the slightest use in quenching the flame—it rather aids than retards the progress of the fire. Earth and sand are the best means of quenching flame.

In Liverpool the authorities have taken advantage of the sandstone formation at the Herculaneum dock to excavate chambers out of the solid rock, which makes a most excellent storage place for barrels. Were tanks constructed within the rock, a safer place for the storage of petroleum could not possibly be devised. At the present time the construction of storage tanks is being pushed on vigorously at Newcastle, Liverpool, London, Barrow ; and in the course of a few years every port of importance will have its petroleum dock refining and storage works. On the Continent this is done extensively at such enterprising ports as Hamburg, Antwerp, Bremen, &c.

The carriage of petroleum (kerosine) in barrels in sailing ships has been carried on for years, and the trade is thoroughly well understood. It says much for the alleged danger of carrying oil in vessels, that fires on board barrel-laden vessels are extremely rare, while during the last two years fires on board cotton-laden ships from the United States have been so common as to demand general attention from owners and underwriters.

23. As regards the question of transport in bulk *versus* transport in casks, it needs no demonstration to see that the balance is in favour of transport in bulk. Of course it always has to be considered that the time may arise when, owing to an increased competition, freights may fall, and thus render the bulk vessel un-remunerative—in which case she is not adapted for any other trade, as is the vessel which simply carries oil either in cases or barrels. Comparing the advantages of carrying oil in bulk as against oil in barrels, Mr. Martell, the chief Surveyor of Lloyds, and an unques-tioned authority on the subject, says, if a steam vessel that could carry 2000 tons of cargo, occupying as a limit 50 cubic feet to the ton, were filled with petro-leum in barrels, she would carry only 1250 tons dead weight. Of this quantity, moreover, about 16 per cent. would represent the tare of the casks, thus re-ducing the actual amount of petroleum to 1050 tons. From this again it is usual to deduct two per cent. for leakage, and if such an amount of leakage does occur in practice, the nett weight of oil usefully carried becomes reduced to about 1030 tons, as against 2000 tons of ordinary deadweight cargo. As the specific gravity of American kerosine is ·80, and that of Russian about ·82, it would occupy in bulk less than 50 cubic feet to the ton, the average volume being about 45 cubic feet, and, therefore, no difficulty in carrying a full cargo would arise. The special fittings requisite for such a method, including the pumps, would, of course, have the effect of greatly reducing

D

the above difference of 970 tons against the barrels, but there would be still left a large margin in favour of the carriage of oil in bulk.

But the advantage due to a difference in carrying power, great though it be, is not the only consideration.   Another important economic question which arises in connection with the barrel system is that of the cost of the barrels themselves.   Their value in the United States is stated to be from 4s. 6d. to 5s. 6d. each, and, with the exception of the few that are taken back to America, they are sold in London when empty for from 3s. 6d. to 4s. each.   The depreciation of from 1s. to 1s. 6d. in the value of the barrel, which amounts to as much as from 350l. to 475l. for one voyage in the case quoted above, is saved under the bulk system.   At present it has to be borne by the consumer of the oil, and, by enhancing the cost, has an effect in restricting the use of the commodity, which would disappear if the employment of casks were discontinued.

24. There is one point in connection with carrying oil in barrels in sailing ships that deserves a passing notice.   In an inquiry as to the collision between an oil-laden vessel and a steamer resulting in the loss of the latter and stranding of the former, it was stated by the captain of the sailing vessel that mineral oil cargoes affected the compass.   In this extraordinary statement he was supported by the evidence of another captain.   It is hardly possible to believe in this age of examinations and technical education, that

a professional man would be found to make such an unwarranted assertion. It is needless to state that mineral oil does not affect a ship's compass. The writer, before giving this contradiction publicity, asked Sir W. Thomson the question, and was confirmed by his statement that mineral oil has no effect upon a ship's compass. It shows the extraordinary ignorance in technical matters that is as a rule possessed by those who conduct our Board of Trade inquiries that none of the assessors who were conducting the inquiry were conscious of anything startling in the statement.

Another master mariner, a " practical man " of very great experience, &c., &c., ran his ship ashore a little time ago, and as usual attributed the disaster to an unlooked-for deviation in the compass. This gentleman attributed the loss of his vessel (an iron steamer) to the fact that the previous cargo had been coal, and the present one was wood, the change from coal to wood affecting the compass! Needless to say this extraordinary theory of compass deviation is not supported by any evidence. It really would seem that many shipowners entrust valuable property to men, not on account of what they know, but on account of what they do not know.

25. For purposes of illumination petroleum ranks next to the arc light. By suitable means, as adopted in lighthouses, a very powerful light is obtained. In many respects lighting with petroleum is better and more economical than gas, colza, or incandescent

lamps (electric). For the illumination of large open spaces petroleum, as burnt in the "Lucigen," the Wells lamp, the Doty lamp, is unrivalled. Here again, in spite of Board Schools and missions, there is not, so far as the writer is aware, a single agency in existence that will teach the class which mostly use oil lamps how to trim, light, and extinguish them. We have any amount of long-haired young men who "slum," women who feel a call for "higher things," clergymen with "views," teachers with a "system," all of whom are willing to do anything on the slightest provocation to obtain a little notoriety, and yet the newspapers continually report "death from the explosion of a paraffin lamp." We frequently see in our large towns that peculiar social feature, the inception, blazing forth, and final collapse of a movement that seeks, by teaching the tenets of a false art, to bring sweetness and light into "blind alleys" and "entries." It is a common thing to see an Oxford youth lecture to an audience of working people about the teachings of Darwin or Herbert Spencer ; and if one cares to hear it, it is not difficult to go where the discussion of the "higher criticism" is carried on in garrets by people who regularly extinguish their wretchedly constructed paraffin lamps by blowing down the chimney ! Who ever sees a course of instruction advertised "on the use of paraffin lamps"? Mr. Marvin certainly has done very much, both by lecturing and writing, to educate people up to the proper standard in this

respect. In the meanwhile, thanks to the universal
demand for cheapness, we have cheap lamps rained
upon the working classes; and the consequent cases
of maiming and burning are to be deplored but
are hardly to be prevented till we teach people how
to obtain a good lamp.

The explosions of paraffin lamps have been and are
a fruitful cause of loss of life and property among the
poorer classes. These explosions are caused in most
cases by using an oil of unsafe qualities and bad
illuminating power, and the carelessness of the
majority of people in the proper handling of oil lamps.
It would be an excellent thing if the well-known book
by Mr. Charles Marvin on this subject were made a
text book in Board Schools; and if the scholars were
taught the elementary technics of petroleum lamps,
instead of the principles of high art, it would result in
a distinct advantage to the community.

26. One great cause which has hindered the introduction of petroleum and liquid fuel in this country is
the absurd restrictions that legislation surrounds the
industry with. Just as we fondly imagine that we can
make people moral and sober by an Act of Parliament, so do we persuade ourselves that danger can be
avoided by imposing all sorts of restrictions upon a very
promising industry. The truth being that no amount
of legislation will make ignorant people educated,
careless people cautious, or reckless people wise.
After half a century of the use of coal-gas in our
houses, have we not still with us the householder

and maidservant who *will* strike a match to see
where a leak is? There are plenty of shipmasters
and officers who *will*, on the slightest provocation,
examine an oil-tank with a naked light. It is useless to blame the coal-gas and petroleum, and talk
absurdly about the danger—there is no danger but
that created by ignorance and folly.

At the same time the State cannot see valuable
lives sacrificed, which might be in most cases pre-
served if ordinary precautions were used. The
explosions of vessels carrying petroleum, and the
numberless deaths from defective lamps, &c., reveal
a state of things which calls for redress. It would
seem unadvisable to harass a valuable and increasing
trade with legislative enactments ; but a few simple
measures might be taken by the Petroleum Association which would go far to prevent any disaster.

A few simple facts about petroleum and lamps
might be drawn up by experts, and the municipal
authorities could insist on retail traders exposing
these " Information " sheets in their shops. Lamp-
sellers, too, might find it to their advantage to dis-
tribute them to their customers, so that the latter
could see the conditions necessary to produce a good
light with perfect safety. As regards those who are
engaged in the Marine Transport of Petroleum, an
increase of knowledge is absolutely requisite. The
Marine Department of the Board of Trade might
do much in this direction ; but so long as the educational status of " Extra Masters " is about on a par

with that of a 6th standard Board School boy, we must not be surprised at captains making such extraordinary assertions as petroleum affecting a ship's compass, and the like.

It is not so much legislation that is wanted, as better and more technical education. No one would rejoice more than the writer if the Board of Trade would use the power it possesses to raise the status of the Mercantile Marine; this can only be accomplished by greatly raising the educational standard of the examinations.

The writer is in the position to see a very great deal of the *personnel* of the Mercantile Marine, and he has no hesitation in declaring that one great cause of loss of life and property at sea is to be found in the low standard of education, or what passes for such, possessed by the average shipmasters. For the safe transport of petroleum it is more than advisable, it is absolutely necessary, that the masters and officers of petroleum-carrying vessels should at least be possessed of the knowledge of the elementary principles of its safe carriage.

CHAPTER II.

27. BEFORE proceeding to discuss the merits of the
different types of oil-carrying steamers it will be
necessary to devote some space to the consideration
of the forces which govern the movements of vessels
when floating in still water. It is very necessary to
have an intelligent knowledge of these forces, as,
without it, it is impossible to say beforehand whether
a given vessel is adapted for her purpose or not. It is
a fact which cannot be disputed that much of the loss
of life and property at sea is due to bad design consequent on these forces not being properly considered.

The writer does not wish to reflect upon his own profession, but the evidence in courts of inquiry has proved over and over again that in too many cases ships are designed and loaded often without those responsible having the faintest knowledge of why a ship floats in one position rather than in another.

Up to perhaps ten or fifteen years back, ships, even the finest, were designed in many cases by pure rule of thumb. Any retired tradesman who wished to dabble in steamship-owning thought himself competent to fix the dimensions and design of a vessel. There were builders who cheerfully undertook to build anything, even the Register Societies knew absolutely nothing of the composition of the materials they dealt with. By a very funny association of ideas it was generally thought that the man who had been a successful shipmaster had an insight into the laws of steamship propulsion, and was therefore the best man to design a vessel. The result of this idea on the part of builders, owners and others was to be found in the many fearful and wonderful examples of what we might term "Primitive Naval Architecture." Most of these vessels are now resting peacefully at the bottom of the ocean. Within the last ten or fifteen years, however, a most marked improvement has taken place. Naval architecture is no longer a rule of thumb process. The investigations and labours of such men as the late Mr. Froude, the late Professor Rankine, the late Dr. J. Woolley, and Canon Moseley, Sir E. J. Reed, W. Denny, M. Daymard, W. H.

White, B. Martell, Professor Jenkins, Professor Elgar, Mr. McFarlane Gray, and others, have been of the most valuable description, and mercantile Jack owes more of his safety at sea to these men than he does to the whole 105 Acts of Parliament which have been passed in favour of this much-legislated-for individual. We can now, thanks to these original investigators, measure accurately the forces acting upon a ship at any given time. We can predict her behaviour. Knowing the stresses and strains she has to encounter, we can provide the necessary strength. Lloyds' Surveyors are no longer a body of men making rules for material they know nothing of. On the contrary, at the present time material passed by Lloyds' Surveyors is a guarantee that it is absolutely reliable and will do all that is claimed for it. One of the results of this immense improvement in ship construction is that we can now carry with perfect safety such cargoes as petroleum in bulk, which was formerly deemed impossible.

Naval architects, engineers and shipbuilders may however, produce a perfect vessel—she may be passed by Lloyds, and stand before the world as perfect in every way—but unless those to whose charge she is committed have at any rate some technical knowledge of the ship and her characteristics, and can apply in an intelligent way the principles of her construction, disaster is very likely, and in fact often does overtake the vessel. The *Ville de Calais* was an extremely good specimen of an oil-carrying vessel. She was lost through the ignorance of those on board, just as the *Austral* was sunk in Sydney Harbour through the culpable ignorance and carelessness of her officers. In this chapter the writer endeavours, in perhaps a very elementary manner, to place before captains and officers of petroleum vessels some of the principles of naval architecture, in the hope that they will be stimulated to study the subject.

28. The writer has avoided as much as possible the use of terms intelligible enough to the draughtsman but hardly so to the sailor. For instance, the expression "moment of inertia of the water-line plane" is unintelligible for ordinary sea-faring people.

However, it is impossible to do altogether without having recourse to mathematical formulæ and expressions. On this subject that eminent authority Sir E. J. Reed says, "It is impossible to carry an exposition of the fundamental principles of stability to any great length without the resort to mathematical expressions." Than the primary principles upon which all such expressions and all stability diagrams depend, nothing can be simpler. The whole weight of the ship and all on board tends downwards under the attraction of gravitation virtually acting through its centre of gravity. The whole buoyancy of the ship acts upwards through its centre of buoyancy. If these two great aggregate forces act in the same vertical line there will be equilibrium. If they act in different vertical lines rotation must ensue, and will continue until the

two lines come together and coincide. The distance
between these two lines when they do not coincide is
the measure of the leverage with which the ship tends
to upright herself or further incline, and whether she
will continue to incline or will return to the upright
depends upon the direction in which the two forces
tend to turn her, which direction is always pretty
obvious. These are really and truly the only essential
doctrines of the stability of ships. It is when you
come to measure the separation of the two lines afore-
said for any given position of the ship, that all the
difficulty and complication enters, because then you
have to take into account the varying form of the
ship, which changes more or less from point to point,
and is comprised within rounded or curved surfaces
the volumes of which it is difficult to measure.

29. When a body floats in a liquid it displaces a
quantity of that liquid equal to it in weight. This is
a problem in Hydrostatics that admits of very simple
proof. As we are not dealing with any other bodies
than ships, and no other liquid than water, we will
confine ourselves to them.

Let us take a vessel and fill it about half full of water
up to the line A B, Fig. 2. Let us now place a floating
body E in the water—this floating body will represent
the ship. We now see the water rises up to the line
marked C D. Take out the body E. It is obvious
that putting E into the water has caused the water to
rise from A B to C D. Suppose we know the weight
of the water in the vessel, up to the line A B, let us

now pour in water till it reaches the line C D, and
then weigh the quantity.   It will be easily seen that
the difference between the two weights will be the
weight of the layer of water A B, D C.

Let us now weigh our floating body E, and we
shall find that it exactly equals the weight of the
layer of water A B, C D ; therefore when a body floats
in a liquid it displaces a quantity of that liquid equal
to it in weight.

When a body is immersed in a liquid it displaces a quantity of that liquid equal to the body
in volume.   This is so self-evident that it requires

FIG. 2.

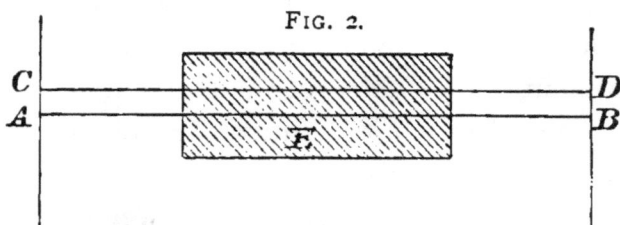

no demonstration.   When a body sinks in a liquid it
is because the weight of the body is greater than the
weight of the displaced liquid.

Example : If we take a piece of lead that contains
10 cubic inches, and which weighs 4·1 lbs., and
measure it in water, it evidently displaces exactly
10 cubic inches of water ; but 10 cubic inches of fresh
water weigh only ·36 lbs., and the lead, in obedience
to the force of gravity acting on it, will sink to the
bottom.   We can put this in another way, and say
that the downward force of gravity is greater in this

case than the upward pressure or buoyancy of the
water.

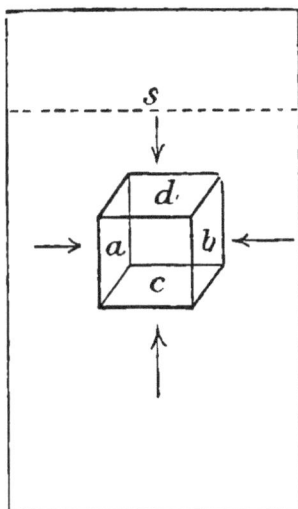

When a body is immersed in a liquid it is pressed
upwards with a pressure equal to the weight of the
volume of liquid that it displaces. To prove this, let
us imagine the cube $a\,b\,c\,d$, Fig. 3, to be immersed in
water. It can be seen that
all four sides will be subjected to the pressure of
the water. The pressures
acting on the sides a and b
will be equal and opposite,
and will neutralise each
other. The pressure on the
side d will be the pressure
of a column of water acting
downwards, whose base is
the side d and height $d\,s$.

The pressure on the side c,
will be the pressure of a
column of water acting upwards, whose base will be
the side c and height $c\,s$. This latter column is greater
than the first one, and so exercises a greater pressure.
The cube will be urged upwards by a force equal to
the difference between the two pressures. This difference is equal to the weight of a column of water of the
same dimensions as the cube, whence it is proved that
any body immersed in a liquid is pressed upwards
by a pressure equal to the weight of the volume of
liquid that it displaces.

When a body is immersed in a liquid it loses a part of its weight equal to the weight of the displaced liquid.

We can prove this by referring to our piece of lead that has a volume of 10 cubic inches and weighs 4·1 lbs. When it is immersed we have seen that it displaces exactly 10 cubic inches of water, which weighs ·36 lbs. If we weigh the lead *when it is thus* immersed its weight will be 4·1 − ·36 = 3·74 lbs.

We can summarise the foregoing thus :—

(1) If a body immersed is of the same density as the liquid the weight, of the liquid displaced being the same as that of the body, the force of buoyancy that tends to raise it is exactly equal to the force with which gravity tends to sink it. The two forces being in equilibrium, the body remains in suspension in any position in the liquid.

A water-logged vessel, that just remains near the surface of the water, may be taken as an example.

(2) If the body immersed is denser than the liquid, it sinks, because the force of gravity urging the body downwards is greater than the force of buoyancy urging it upwards. Referring again to our piece of lead, we see that lead being much denser than water it therefore sinks.

(3) Lastly, if the immersed body is lighter than the liquid, the buoyancy prevails, and the body rises until it displaces a quantity of the liquid equal to itself in weight—it is then said to float.

Applying the foregoing principles to actual ships.

we see that when a ship is floating she displaces a quantity of water equal to herself in weight, and when she sinks it is because her weight, from some cause, becomes greater than the volume of water she displaces.

30. When a vessel floats at rest in still water she is subject to two equal but opposite forces. There is the weight of the vessel, which, under the influence of the force of gravity, continually urges the vessel downwards. There is also the force of buoyancy, acting upwards. Should the force due to the weight of the ship and cargo exceed the upward pressure of buoyancy the ship sinks. Should the two forces be equal the vessel floats. The centre of gravity (C.G.) is that point in a body on which the body will balance in all positions, or, for practical purposes, we may say that the C.G. of a ship is a point within her at which the whole weight of the ship acts.

The centre of buoyancy (C.B.) is the centre of gravity of the displaced water, or, what is the same thing, it is the centre of gravity of the space occupied by that part of the ship which is wetted by the water. It is a point within a ship at which the whole upward force of buoyancy may be said to act.

Thus the vessel (see Fig. 4) is floating at rest and upright. G is the position of the C.G., and B is the position of the C.B. These two forces balance each other. A ship, being symmetrically built, and, we may suppose, properly laden, will have the same amount of weight on each side of the middle line, hence the C.S.

will be on the middle fore and aft line.   So long as the ship is upright the C.B. will likewise be, on the middle fore and aft line.

31. Let us now cause the ship to heel over to a small angle by the operation of some external force,

FIG. 4.

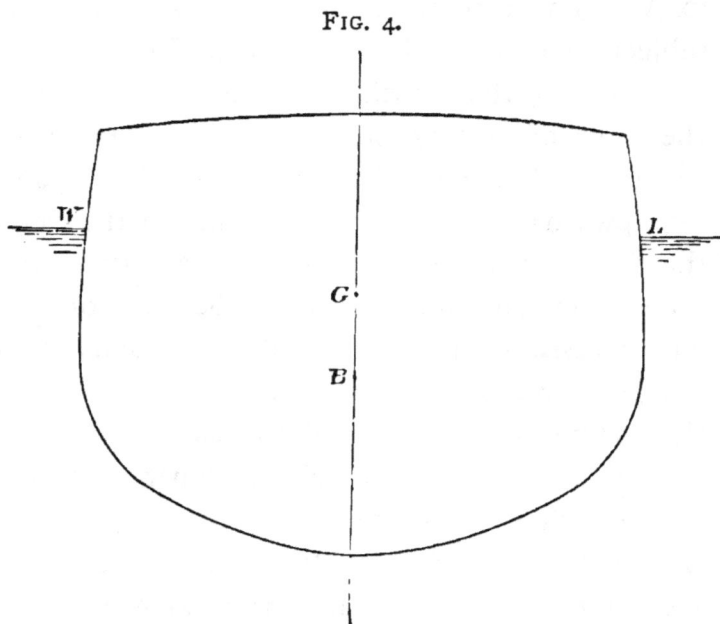

as the wind.   It will be noticed (see Fig. 5) that the water-line is altered.   By heeling the ship to starboard we have caused a volume of the ship denoted by W' O W to emerge on the port side, and a corresponding volume, L' O L, to be immersed on the starboard side ; and the volume of the wetted portion of the ship is now that below the line W' O L', which is of a different shape to W O L.   It will be obvious that the C.G. of the two shapes will not occupy the same spot.   In the

E

upright condition we have seen that B represents the
position of the C.G. of the immersed part of the ship,
or is the centre of buoyancy, but we find that B' is
now the centre of buoyancy or the C.G. of the immersed part of the ship.   The distance B B' is found
by applying the following principle in mechanics.

FIG. 5.

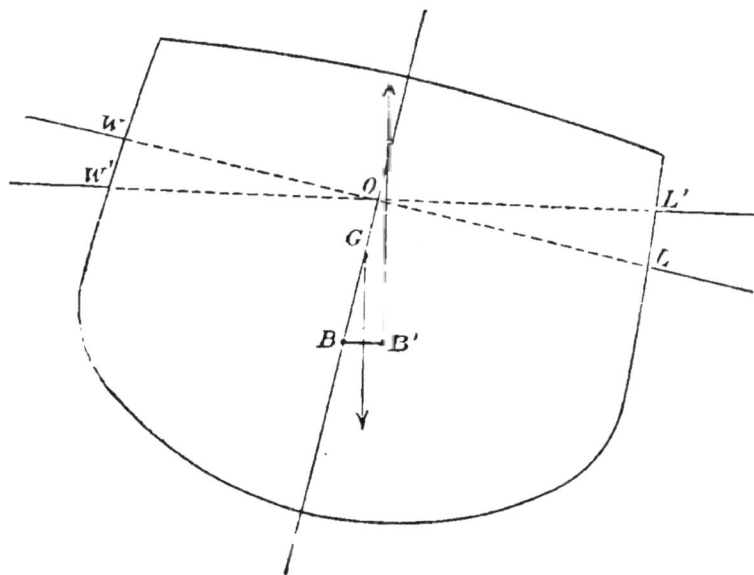

32. Let A B C D be a body (Fig. 6) with a weight
W whose centre of gravity is G, and let the figure of
this body be altered by transposing a part whose
weight is w_1 from the position E C F to the position
F D H, so that the new figure is A B H E.   Let g_1
be the original and g_2 the new position of the centre
of gravity of the transposed part, then the centre of
gravity of the whole body will now be at G_1, and the

distance $G G_1$ will be on a line parallel to g_1, g_2 and equal to $\dfrac{g_1 g_2 \times w_1}{W}$.

33. Applying this to the ship we see that B B', Fig. 5, will be on a line parrallel to the line joining the centres of gravity of the volumes W' O W and L' O L, and the length of B B' will be found by multiplying the length of the line joining the two C.G.'s of the volumes W" O W, and L' O L by one of these volumes, and dividing by the total volume of the ship.

FIG. 6.

This is what is substantially done in ship calculation. Turning back to Fig. 5 we have then the weight of the ship acting downwards in a vertical line from G and the force of buoyancy acting up through B'. These forces are equal and opposite. A little consideration will show that the result will be to turn the ship back again to the upright condition. This is an example of stable equilibrium.

34. The force of gravity acting downwards, and the force of buoyancy acting upwards, and these two forces, not meeting in the same vertical line, form what is known in mechanics as a "couple."

A "couple" may be defined as consisting of two

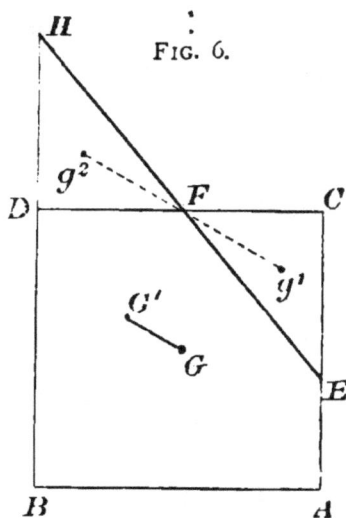

parallel forces which are equal but act in opposite directions. The arm of the couple is the perpendicular distance between the lines of action of its forces. The moment of a couple is the product of one of the equal forces into which the arm, that is, the number which expresses the force, must be multiplied by the number which expresses the arm, to produce the moment.

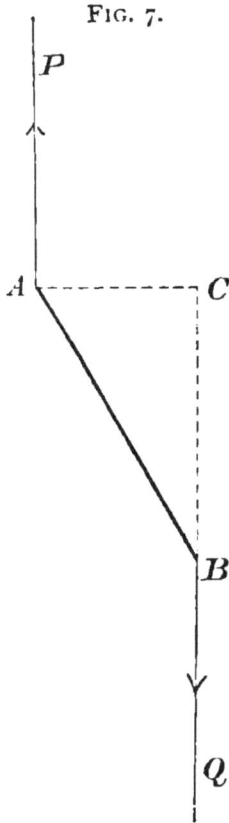

FIG. 7.

In Fig. 7 let A B be a rigid bar, and let a force P act vertically upwards at A, and a similar force Q act downwards at B, the result will be to turn the bar till it lies in the same vertical line as the forces, and the amount of this force turning the bar is found by multiplying one force by the perpendicular distance between the two forces. Thus we should have, moment to turn the bar is P × A C, but $\dfrac{AC}{AB} = \sin A\,B\,C$, therefore the moment tending to turn the bar is P × A B sin A B C.

35. Applying the foregoing principle to an actual ship, we have, Fig. 8, the weight of the ship, or displacement in tons = D, acting down through G.

We have the buoyancy acting up through B, and cutting the vertical midship line at a point M. M G we may regard as a rod acted on at its ends by these two forces, the moment of the couple will be D × G Z (the perpendicular distance between the two forces), but G Z = G M sin a (the angle of inclination), therefore

FIG. 8.

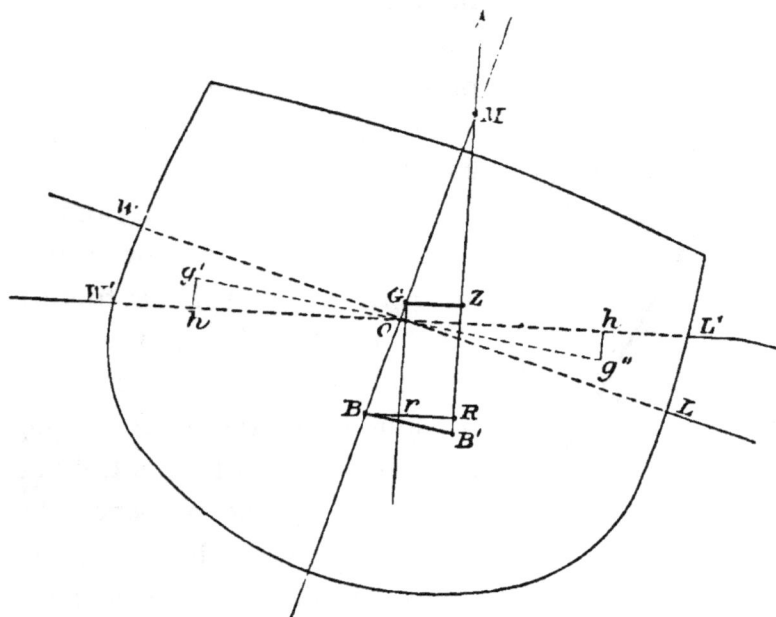

the moment of the couple becomes D × G M sin a. We have seen that the distance B B' is determined by the

formula $B B' = \dfrac{v \times g' g''}{V}$.

Where v is the volume of one of the wedges, and $g' g''$ is the distance between the centres of gravity of the wedges of emersion and immersion, and V is the

volume of the ship, instead of using the distance
$g' g''$, we might use the distance $h h_1$, which is
measured along the L W L. From B draw B R
parallel to G Z. B R represents the distance that B
has shifted, measured in a direction parallel to the
L W L.

$$B r = B G \sin a.$$

36. We now have

$$G Z = r R = B R - B r = B R - B G \sin a,$$

but,

$$B R = \frac{v \times h h'}{V},$$

therefore,

$$G Z = \frac{v \times h h}{V} - B G \sin a.$$

This is the fundamental formula for the statical
stability of a ship; it is know as Atwood's formula.

This represents the leverage with which the weight
of the ship acts, and the arm G Z is frequently spoken
of as the stability lever.   G Z alters its value for each
angle of keel.

37. We will now just notice the case of a vessel that
is unstable.   We often see a vessel when light with
a list of 4 or 5 degrees.   This does not, as is sometimes supposed, indicate that she is therefore unstable,
it merely says that she has more weight on one side
than the other, and consequently will heel over till
she finds a position of stable equilibrium.   When
upright, owing to the arrangement of the machinery
or disposition of the cargo, the C.G. may not be in the

middle line, but a little on one side. She will heel over till the vertical lines passing through G and B are in the same vertical plane, when there will be equilibrium. If the C.B. moves out, so as to always be under C.G., the vessel will float in any position.

38. Many vessels, by reason of peculiar design or bad disposition of weights, will float with perfect safety in almost any position, and a weight of a few bales of Manchester goods at the end of a derrick will often be sufficient to produce an inclination of several degrees. When this occurs it shows that M and G occupy practically the same spot till sufficient weight is put below, when the C.G. falls. It some-times happens that a vessel, when laden with an homogeneous cargo, will encounter a gale during which the cargo shifts. The cargo becomes denser on the lee side and a vacant space is left on the weather side—such a case is shown in Fig. 9. When upright the C.G. of the ship and cargo is at G ; supposing she were inclined, and the cargo did not shift, she has a metacentric height shown by G M. The cargo, how-ever, does shift, and the C.G. is at G', and the C.B. at B. The result is that a turning force equal to the weight of the ship multiplied into the arm M Z is introduced, and the ship gradually heels over more and more till she ultimately founders. In this case we have not considered the forces introduced by the heave of the sea. It shows how very necessary it is to have the most careful stowage in such vessels. Numbers of such vessels lie at the bottom of the sea.

Courts of inquiry are held, and any cause but the right one is generally put forward to account for the loss.   The court itself, composed as it usually is of a magistrate, whose nautical knowledge is gained by a few "personally conducted" trips, one or two ship-masters, no doubt eminent seamen, but possessing

FIG. 9.

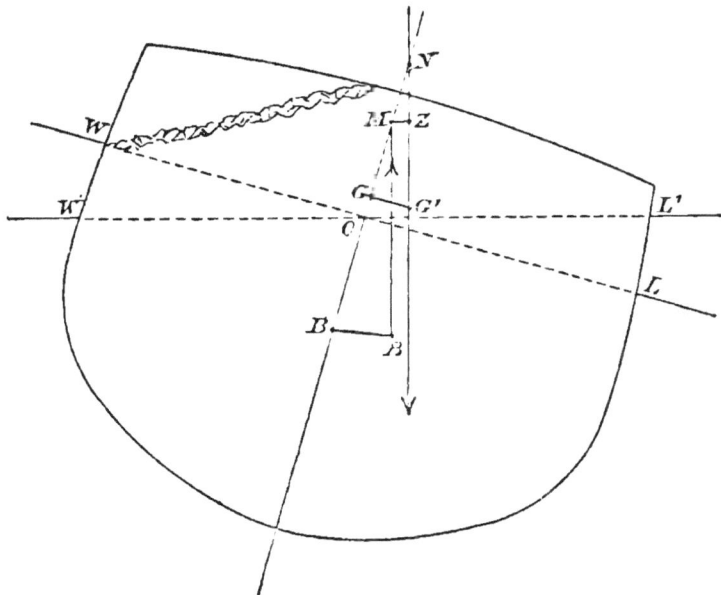

little or no technical knowledge of the science of stability, find the unfortunate shipmaster to blame when really the blame in too many cases lies with the designer of the ship.   We place the name of the architect who designs a church on the foundation stone.   Why should we not place the name of the individual who is responsible for the design of the

ship on all the official documents relating to her? so
that in case of loss some one will be held directly
responsible?

39. Turning back to Fig. 8, it will be noticed that
the vertical line passing through the C.B. continued
upwards, intersects the vertical midship line (produced) of the ship at the point M. This point is
called the "metacentre." If, starting from the upright position, we gradually heeled our vessel through
very small angles up to say 10°, we should find
that each position of the C.B. would correspond to a
certain point very *near* the vertical midship line, and
that a certain relation obtains between these points
of intersection of successive axes, and the curve of
centres of buoyancy. In an elementary sketch like
the present we need not do more in this relation than
to say, that the curve traced out by the intersection
of successive axes is called a curve of pro-meta-
centres.

For our present purpose it will suffice if we say
that the metacentre is a point in the vertical midship
line of the vessel, and its position is found by drawing
a vertical line through the C.B. This point (the
metacentre) is of very great importance.

In Fig. 8, we have a strong righting force in action,
it will be noticed that M is above G. In Fig. 10
we have a capsizing force, and M is below G. The
height G M is called the "metacentric height." When
M is above G this metacentric height is called positive; when below, as in Fig. 10, it is called negative.

When there is positive metacentric height it indicates
by its amount the stability of the ship.

40. The position of the metacentre derives its
great importance from defining the limit to which the
C.G. can be raised without introducing instability.
When the ship is upright, and for small angles of

FIG. 10.

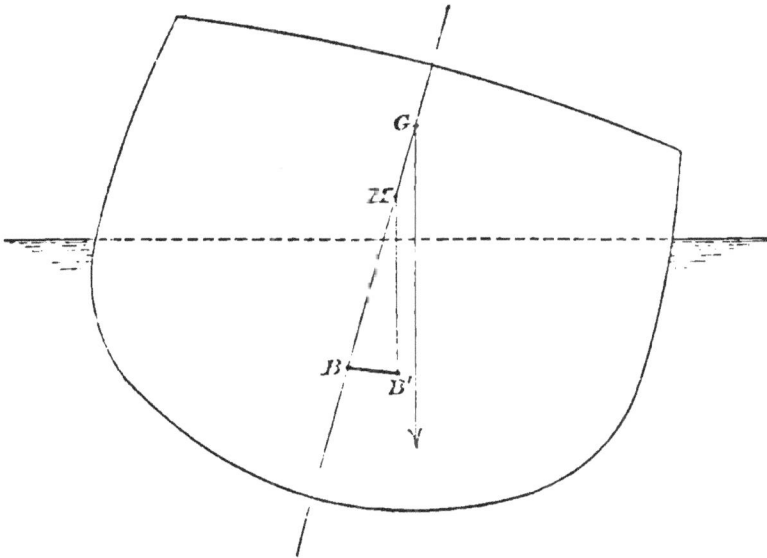

heel, the position of M indicates the initial stability of
the ship. If the points M and G lie very close together
it indicates that the vessel will be tender. If M is
below G the vessel is crank and will "loll" from side
to side by merely shifting a few small weights. If G
is much below M the vessel is stiff. If it is very
much below the vessel will be too stiff.

41. To judge of the stability of a ship at large

inclinations we require to know the values of the lengths of G Z. Knowing the length of G Z for a few angles, we can construct a curve of stability that tells us the value of this righting force at any given angle.

42. If, from an inspection of the curve of stability in a given condition of the vessel, it is found that she has too little righting power, measures can be taken to remedy this before getting to sea. It is much better and safer to know all about the stability of the ship, and act accordingly when in port, than to wait till the vessel is at sea to discover whether she is tender or not. The number of vessels that load homogeneous cargoes, and put to sea with all the confidence in the world, only to find too late that the ship is tender, is enormous. Captains very often forget that the consumption of coals and water in most cases diminishes the vessel's stability. Often too, when at sea, and it is discovered that the vessel shows a remarkable disposition to sail on her side, a remedy is found in filling up the ballast tanks. This is always a most dangerous proceeding, as will be shown later on ; during the operation the vessel loses stability rapidly Apart from this a great strain is brought upon the tanks which may result in damage to the cargo. If, after "running up the tanks," the sea cocks are left open, a pressure will be brought upon the tanks that, unless the vessel is exceedingly well constructed, is bound to result in leakage. The upshot usually is that the captain has his services dispensed with.

A common idea is that the consumption of coals and

stores increases a vessel's stability, whereas in ninety-nine out of a hundred vessels it is just the reverse. The captain says, "I am burning so many tons per day; my ship is *increasing her freeboard*, and is therefore more seaworthy." To his surprise his vessel becomes more tender and evinces a tendency to sail on her side; bad weather is met with, and it is sought to obtain stability by filling a tank. As the writer says, how many vessels have suddenly turned turtle during the operation is unknown, but the number must be very large. Seamen, in most cases, blindly pin their faith on freeboard, but freeboard is, after all, but one and not the most important factor in the seaworthiness of a vessel.

All this vexation of spirit could be avoided if it were clearly understood what constitutes a stable vessel. There are troubles connected with navigation that are quite enough to bear without having any anxiety as to the behaviour of the vessel in bad weather. A deficiency of stability can in general be best made good by leaving the tween-decks empty, or only partially filled. This, of course, means shutting out cargo. On the other hand, if the tween-decks are kept filled, and stability is assured by keeping the ballast tanks filled, a great increase of draught is the result. This is, however, in the writer's opinion, not to be feared, provided always that the upper deck can be made practically a water-tight flat. Few vessels have their hatches and deck openings so constructed as to be impervious to water; and yet there can be

no question that vessels intended to carry the most
cargo on certain dimensions, and with the minimum
of freeboard, should have their upper deck capable
of being made a water-tight flat.   The writer sees no
reason why this should not be done.

43. So far we have defined the points B, G, and M,
but only for one draught of water of the ship ; the
point B obviously alters with each change in draught.
The point G is fixed, so far as the hull of the ship is
concerned, but with cargoes of varying density and
amount it will either rise or fall.   The point M, being
dependent upon the C.B., will be determined by the
draught.

To find the position of M is a very simple thing
with bodies of constant cross-section, such as rectangles
cylinders, and wedge-shaped bodies, but a ship being
a body of very varying dimensions the problem is
complicated.*   In any vessel

$$B M = \frac{\text{Moment of inertia of water-line plane}}{\text{Volume of displacement}}$$

The expression "moment of inertia," merely means
that we take the mass of every particle of a body,
and multiply it into the square of its distance from
an assigned straight line.

---

* The general expression for the height of the metacentre above the
centre of buoyancy is

$$B M = \tfrac{2}{3} \int y^3 \frac{d\,x}{V},$$

where y is the half breadth of the water-line plane, d and x the
values of small finite dimensions.

44. As we frequently speak of "curves" it may be well to say a few words as to their meaning. It is well known that results of calculation, when given in a tabular form, merely record certain facts ; the law which underlies these is known to the calculator, but the object is to convey the consequence of the law to the minds of people who in most cases have not had the necessary training to enable them to arrive at the law itself. To convey a series of results in an intelligible manner and without the aid of mathematical formulæ a graphic method is used. Nautical people are familiar with this principle as exemplified in Napier's diagram of compass deviation, by means of which the changes in the deviation are conveyed to the mind in a manner that would be impossible were mathematical formulæ used.

When two series of values depend on each other they can be represented either by means of a pair of comparative scales or by means of a curve. Referring to the displacement scale for example, Fig. 11, each point on it represents two quantities, each quantity belonging to one of the series of values which depend upon one another. The one quantity is measured by the position of the point as read on the horizontal scale, the other by its position as read on the vertical scale to the right or left of the diagram (the former of these two measurements being called the abscissa and the latter the ordinate of the point). If one of these two quantities, say the abscissa, is known, the other, the ordinate, can readily be found by setting off the part

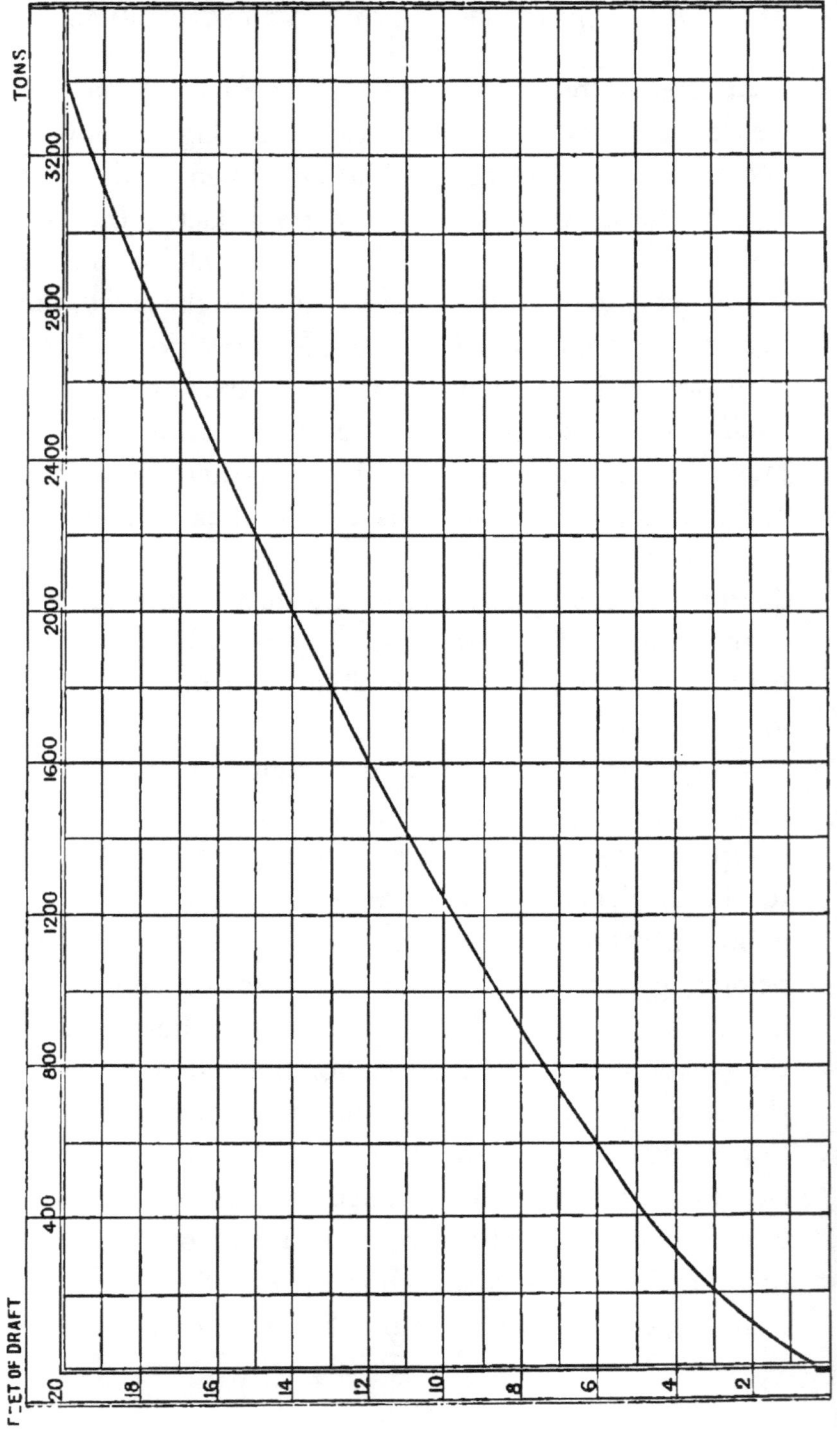

FIG. 11.

on the horizontal scale and drawing a line parallel to
the scale-line on the left, until it cuts the curve.   The
distance of the point of intersection above the base-
line measured by the scale on the left is the quantity
required.   If, on the other hand, the ordinate is
known and the abscissa required, the process has
simply to be reversed.

45. When a ship is floating in still water we have
said that she displaces a quantity of water equal to
herself in *weight*. When she sinks she displaces a
quantity of water equal to herself in *volume*. If we
calculate the total volume of the vessel and divide by
35 we shall have the weight of the water that the ship
displaces when she is just at the point of sinking.
When a vessel is floating at her load draught she will
generally displace from 60 to 75 per cent. of total
volume of the vessel ; the remaining 40 or 25 per cent.
is called the reserve, or surplus buoyancy, and represents the extra weight that must be put on board to
sink the ship. Example :—A rectangular prism,
300 × 40 × 25, floats at a draught of 20 feet. Displacement of vessel when just floating is

$$\frac{300 \times 40 \times 25}{35} = 8571 \cdot 43 \text{ tons} ;$$

when floating at a draught of 20 feet, it is

$$\frac{300 \times 40 + 20}{35} = 6857 \cdot 14 \text{ tons} ;$$

$$\begin{array}{r} 857 \bar{} \cdot 43 \\ 6857 \cdot 14 \\ \hline 1714 \cdot 27 \end{array}$$

The surplus buoyancy will be

$$\frac{1714\cdot3 \times 100}{8571\cdot4} = 20 \text{ per cent.,}$$

or in other words, such a vessel would have to take on board a weight of 1714·3 tons before her upper deck would be level with the water.

46. Displacement Scale.—By means of this we can at any time tell the weight of the cargo on board. When accurately constructed it is of the very greatest use on board ship, and furnishes indisputable evidence as to the weight which is taken in or discharged. A vessel floating at rest displaces a volume of water equal to her own weight. Given therefore the volume of the immersed part of the ship, and dividing by the number of cubic feet of water to one ton, we obtain the displacement of the ship in tons. The volume of the ship at each foot of draught is calculated with great accuracy from the drawing of the ship. If the vessel floats in pure fresh water specific gravity 1·000, the volume in cubic feet is divided by 35·905. If she floats in sea water, specific gravity 1·027, the volume is divided by 34·973 or usually by 35. The quotient is the number of tons that the vessel and all on board weighs at that draught. A vertical scale is drawn on which is (Fig. 11) marked, the draught of the vessel in feet ; at right angles to this vertical scale is drawn a scale on which is set off the scale of tons. Against each draught is drawn a line equal to the displacement at that draught as taken from the scale

F

of tons ; the ends of these lines are joined together and
a curve is obtained which is the curve of displacement.
At any given mean draught it is only necessary to
draw a line till it cuts the curve and apply it to the
scale of tons, and the displacement is obtained at
once.

47. Another scale that should be used in conjunction
with the displacement scale is the " curve of tons per
inch," this is constructed in a similar manner to the
displacement scale. A vertical scale (Fig. 12) shows
the draught of the vessel, and at each foot of draught
is drawn a horizontal line showing the number of tons
it takes to immerse the vessel an inch at that draught.
This scale is of great use in checking small weights.

A properly constructed displacement scale and
scale of tons per inch should be on board every
vessel. Owners should, when contracting for a vessel,
stipulate that the builder should guarantee the substantial accuracy of all plans, scales, &c., supplied.
This is not perhaps necessary with the very first class
firms, but it is not unknown that the plans, scales,
&c., supplied do not give accurate results in all cases.

48. It is frequently of service to the intelligent shipmaster to have at his disposal data which are not
usually given. How many shipmasters know the
sectional L.W. area, or the area of the immersed
midship cross section, or the block coefficient
of their vessels ? yet these particulars are often of
great importance, and we will give a few examples
of their practical use.

The area of the L.W. section of the ship is found in the ordinary way by the application of Simpson's Rules. Knowing this area in square feet, we can tell

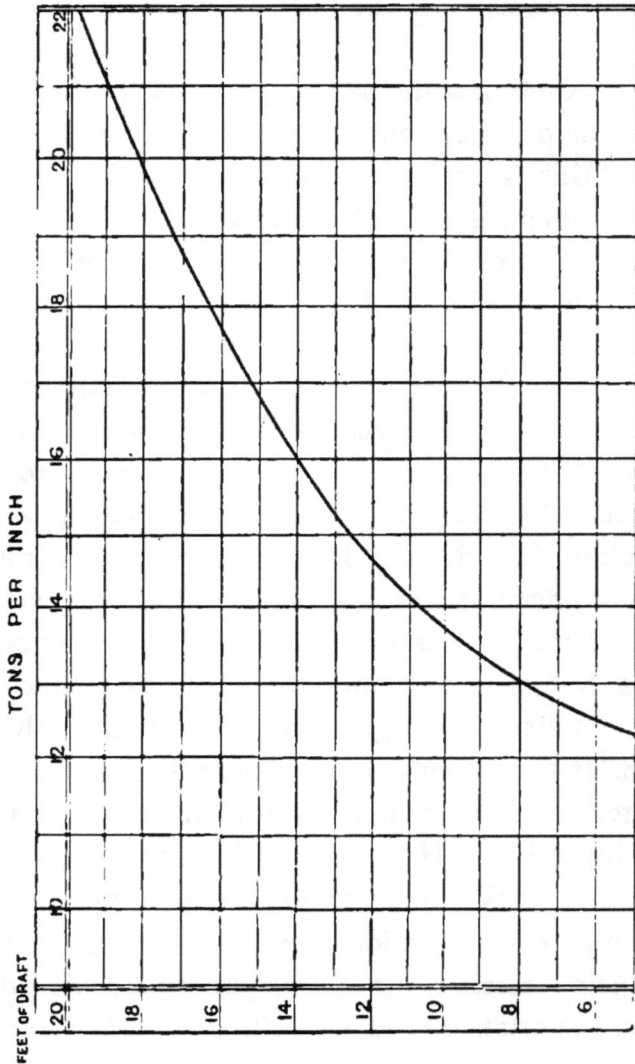

FIG. 12.

TONS PER INCH

FEET OF DRAFT

the amount of sinkage due to putting on board so
many tons of weight. We have said that 35 cubic
feet of sea water equal 1 ton of displacement; that
is, for every ton put on board, the ship will displace
35 more cubic feet of water. If we multiply the
sectional area of the ship by 1 foot, we obtain the
volume in cubic feet of the ship between two water-
lines 1 foot apart.

The area of the load water-line section of a steamer
is 5050 square feet, and she takes on board 300 tons
of bunker coal. How much should she sink in sea
water?

$$\frac{300 \times 35}{5050} = 2 \cdot 08 \text{ feet (nearly)}$$

$$= 2 \text{ feet } 1 \text{ inch (nearly).}$$

This at once gives us a simple means of checking
the coal merchant.

When loading a vessel in sea water, it is found that
she sinks 1 inch for each 25 tons of cargo put on
board. What is her sectional area at that load
water-level?

$$35 \times 25 = 875 \text{ cubic feet of displacement;}$$

$$1 \text{ inch} = \tfrac{1}{12} \text{ of a foot;}$$

$$\frac{875}{\tfrac{1}{12}} = 10,500 \text{ square feet} = \text{sectional area.}$$

49. The sectional area of a ship at the L.W. line is
8540 square feet, her draught was then 21 feet 6 inches
fore and 22 feet 3 inches aft. She takes in a quantity of bunker coal supposed to be 300 tons, and the

coal merchant presents his bill for this amount to the chief engineer for his endorsement. After completing the bunkering the draught is 22 feet 2 inches fore and 23 feet 10 inches aft. Is the engineer justified in putting his signature to the coal bill?

Here we have before bunkering

$$
\begin{array}{ll}
21 \text{ feet } 6 \text{ inches} & \text{fore} \\
22 \quad ,, \quad 3 \quad ,, & \text{aft} \\
\hline
21 \text{ feet } 10\frac{1}{2} \text{ inches mean draught.}
\end{array}
$$

After completion of bunkering we have the draught

$$
\begin{array}{ll}
22 \text{ feet } 2 \text{ inches} & \text{fore} \\
23 \quad ,, \quad 10 \quad ,, & \text{aft} \\
\hline
23 \text{ feet } 0 \text{ inches} & \text{mean draught} \\
21 \quad ,, \quad 10\frac{1}{2} \quad ,, & ,, \qquad ,, \\
\hline
1 \text{ foot } 1\frac{1}{2} \text{ inches.}
\end{array}
$$

The ship has then sunk $13\frac{1}{2}$ inches. To find the tons per inch we have

$$\frac{35 \times x}{\frac{1}{12}} = 8540$$

$x = 20\cdot33$ tons. That is, at this draught every $20\cdot33$ tons sinks the ship 1 inch. Now, if we had received 300 tons exactly, the sinkage would be

$$\frac{35 \times 300}{8540} = 1\cdot23 \text{ feet}$$

$$= 1 \text{ foot } 2\frac{3}{4} \text{ inches,}$$

but the ship actually has only sunk 1 foot 1½ inch.

$$
\begin{array}{cc}
1 & 2\frac{3}{4} \\
1 & 1\frac{1}{2} \\
\hline
 & 1\frac{1}{4}
\end{array}
$$

Now 20·33 × 1·25 = 25·4 tons, clearly then the shipowner would lose the value of this amount of coal, unless the proper precaution of working out a few simple figures was indulged in. "Oh," says the practical man, "I can tell by the look of the bunkers if I have my quantity of coal." Any one who has carefully watched the ingenious methods adopted at Malta, Port Said, and other places, of building up coal and piling it loosely, will know that a ship seldom if ever gets her proper quantity. In the foregoing example, knowing the tons per inch obtained from the scale of tons per inch, or from the above method, we could at once tell the draught the ship should have.

The mean draught before loading was 21 feet 10½ inches.

$$
\frac{300}{20\cdot33} = 14\cdot75 \text{ inches.}
$$

$$
\begin{array}{rcc}
\therefore & 21 & 10\frac{1}{2} \\
+ & 1 & 2\frac{3}{4} \\
\hline
 & 23 & 1\frac{1}{4}
\end{array}
$$

is the mean draught the ship should have after loading.

50. In cases of dispute, the captain should always

ascertain the specific gravity of the water the vessel is floating in. If he has not a proper hydrometer he can make use of the engineer's salinometer. The actual displacement can be obtained by multiplying the displacement as given on the displacement scale by the specific gravity found, and dividing by the specific gravity of sea water. In Fig. 13 is given the relation between specific gravity and cubic feet per ton weight, also the reading on the salinometer in ounces to the gallon, and number of thirtieths of salt contained in the water.

51. It is frequently of service to be able to say accurately (not "about," as they say in courts of inquiry) the draught of a ship when going from water of one density to another. We must know the specific gravity of the water, and

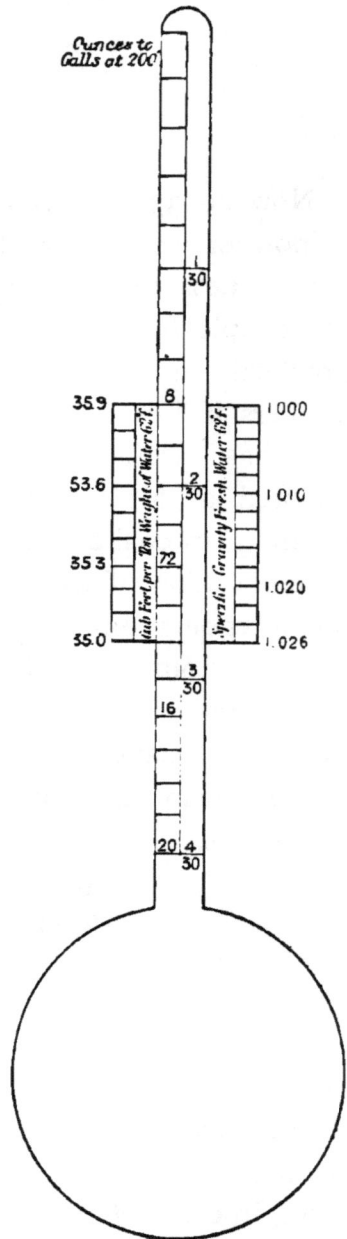

FIG. 13.

the draught of the ship. Below is a list of the specific gravities of the various seas and the weight per cubic foot.

			Sp. gr.	lbs.
River water	..	..	1·000	62½
Sea of Azof	..	..	1·008	63
Black Sea	..	..	1·014	63¾
Baltic	..	..	1·015	63₁⁷₆
Mediterranean	..	..	1·030	64¾
North Atlantic	..	..	1·028	64¼

Then to find the draught becomes a simple matter of proportion.

Example : A vessel draws 21 feet 6 inches off Dungeness, what will be her draught in the London Docks?

$$\begin{array}{cc} \text{ft.} & \text{in.} \\ 21 & 6 = 21\text{·}5 \end{array}$$

$$\frac{21\text{·}5}{1\text{·}000} = \frac{x}{1\text{·}028}$$

$$\begin{array}{ccc} & \text{ft.} & \text{ft.} \quad \text{in.} \\ \therefore \quad x = 22\text{·}1 = & 22 \quad 1\tfrac{1}{4} \end{array}$$

We can solve this question in another way :—

Let W = displacement of ship in tons ;
    d = tons per inch at L.W.L. in sea water ;
    I = increase of draught in river in inches ;

$$I = \frac{W}{63 \times D}$$

52. It can be demonstrated that when a vessel goes from water of one density to another, she also

changes trim. When a ship leaves the river and
proceeds to sea, she trims by the stern, but going
from sea to river she trims by the bow. We do not
discuss this any further, as the amount of change of
trim is so slight as to be practically negligible,
although it is of importance in view of the perplexing
and extraordinary calculations that assessors indulge
in when seeking to arrive at the amount of freeboard
that a vessel had, and which at the time of the inquiry
is at the bottom of the sea. So far as the writer is
aware, the question of change of trim has never arisen
in the courts.

53. Curve of righting levers or curve of statical
stability.—We have seen that the force tending to
bring a ship to the upright condition is measured by
the length of the arm G Z. The value of G Z alters
for each inclination of the vessel. If we calculate the
values of G Z for each successive inclination of the
ship from the upright 0°, to when she is on her beam
ends 90°, and plot out these values as a curve, we
have a graphic form which enables us to tell the
amount of stability for any intermediate angle, but
only for that particular draught of water. We can
combine curves of stability for different draughts by
means of what are termed cross curves of stability,
and by their aid we can tell the righting force in any
condition of the ship.

To construct an ordinary curve of stability, a
vertical line is drawn, on which are spaced off feet
and tenths. A horizontal line is drawn, divided into

degrees (Fig. 14). At angles of 10°, 20°, 30°, &c., the
lengths of the righting levers are set up, and a curve
drawn through their extremities.

Instead of expressing the value of the righting
levers in feet, it is sometimes convenient to measure
the *moment* of the stability, that is, the product of
the length into the weight, and is obtained by multiplying the length of the righting lever G Z by the
displacement of the ship in tons, and the force is
thus expressed in foot-tons. Suppose we raise
2000 tons 5 feet high, we do 10,000 foot-tons of

FIG. 14.

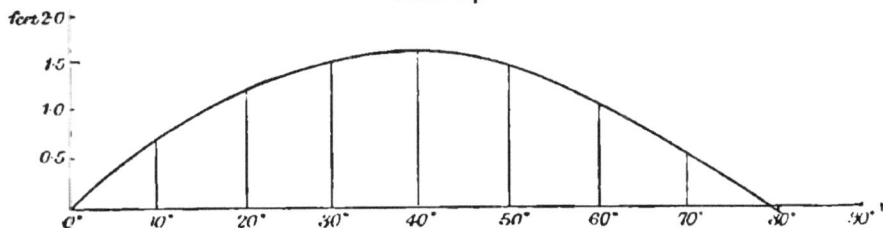

work, which is the same as raising 1000 tons 10 feet
high, or raising 10 tons to a height of 1000 feet.
Suppose a ship of 1000 tons displacement is heeled
over to an angle of 10° by the wind, at which angle
the value of G Z is ·75 feet. The numeral of the
force producing this inclination is

$$D \times G Z$$
$$1000 \times ·75 = 750 \text{ foot tons.}$$

54. It will be noticed that the curve starting from
0° gradually ascends till it reaches a maximum. The

ordinate at this point represents the angle of maximum stability. The curve then descends till it cuts the base line at 78°. The latter point is called the angle of vanishing stability, because if the vessel was inclined to the angle, she would have no tendency to right herself. The number of degrees included between 0° and the vanishing point is called the range of stability.

55. The Curve of Metacentres and Buoyancy.—We have seen that the position of the metacentre alters, being dependent upon the position of B. It is usual to combine the position of the metacentre and centre of buoyancy corresponding to it on one curve. Having obtained the position of B and M for a number of different draughts, a curve is drawn through each set of points, and we are thus enabled to tell the height of the metacentre above the C.B. for any intermediate draught. Such a curve, called a metacentric diagram, is constructed as follows (Fig. 15) :—

A vertical scale of feet of draught is drawn, and from one extremity an oblique line is drawn at an angle of 45° from the points at which this oblique line intersects the water-lines—vertical lines are drawn, upon which are set off the distances down of the centres of buoyancy and below their corresponding water-lines, and from these centres of buoyancy are set up the corresponding heights of the metacentres. A curve is passed through each set of points thus obtained, and we thus have a curve of centres of buoyancy, and a curve of metacentres.

56. There are other methods of constructing and
arranging these curves, but the one we have given is
perhaps the simplest and best adapted for practice.
The great value of a metacentric diagram is the facility

FIG. 15.

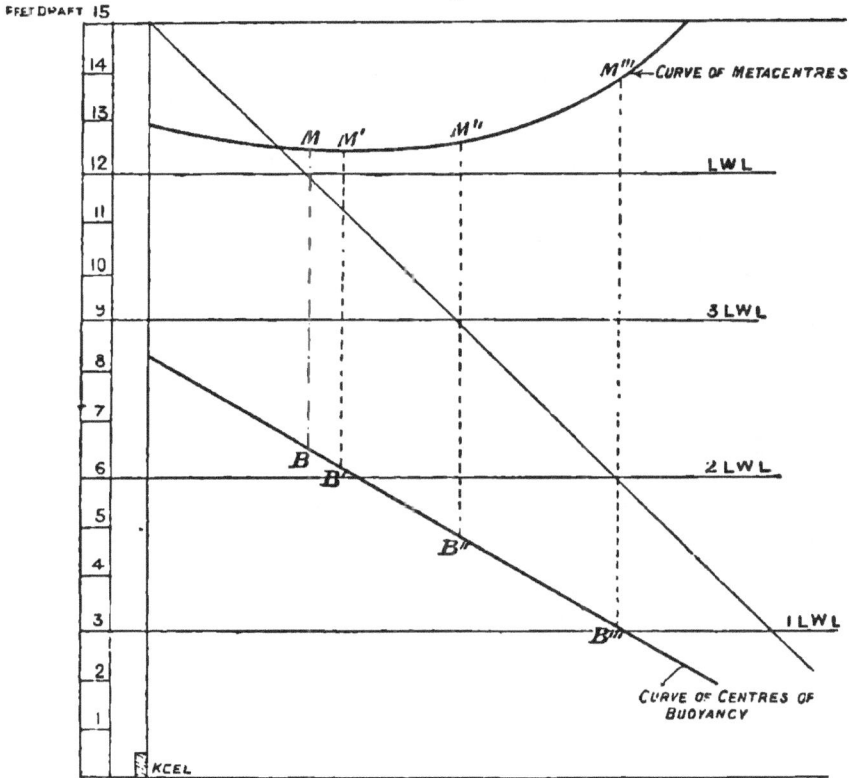

it affords for indicating the stability of the ship at
various draughts when the position of the centre of
gravity is known. It must, however, be used in conjunction with the curve of stability, because it is quite
possible to have a good amount of metacentric height

associated with small lengths of righting levers and
a small range of stability.   In fact, it is not enough
for the captain to know that he has a good amount
of metacentric height.   He also wants to know the
amount and range of stability as shown by the curve
of righting levers.

57. We have seen that the value of the righting
force is shown by the curve of stability, but this curve
only holds good for the particular conditions under
which it is calculated.   If we get out the curves of
stability for the ship, say in the extremely light condition (no cargo, no ballast), and floating at the least
draught ; then with ballast in and bunker coals,
but no cargo ; then with the ship fully loaded, in
proper trim, &c. ; and finally a curve when the ship is
loaded with either a very light or very heavy kind of
cargo ; we shall have a very useful mass of information that, properly used, will save many an
anxious hour to the captain.   In making use of a
curve of stability, it must be carefully noticed
whether the conditions are satisfied under which that
particular curve is constructed.   Stability is often
obtained by careful designing (which, by the way,
is not a feature of all cargo boats).   We have what
is termed "stability due to form."   Without enlarging
on this, we may easily see that in two vessels having
the same freeboard, the one with most beam will
have, other things being equal, the greatest amount
of initial stability ; but her range is limited at a
comparatively small angle, whereas the narrow vessel

will naturally be less stable, but will have a larger range.

58. We will now give a few examples of the stability of cargo steamers, taken by permission from the invaluable book by Sir E. J. Reed, M.P., entitled 'Stability of Ships'—a book which should be found in every vessel, from the mail steamer to the collier. The writer takes this opportunity of testifying to the great practical advantage he has derived from studying it.

The Fig. 16 represents the curves of stability for a

FIG. 16.

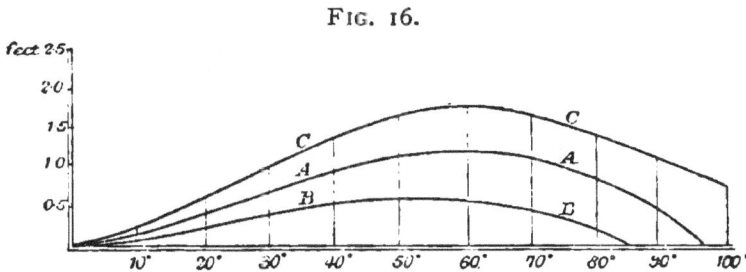

raised quarter-deck steamship of the wall-decked type, dimensions 267·5 feet by 35·5 feet by 19·7 feet. Tonnage under deck 1509, gross 1866, net register 1009. Curve A represents the ship laden with a homogeneous cargo, which entirely fills the hold, the bunkers being assumed as quite full of coal, the boilers filled with water and all stores on board, but ballast tanks empty. The displacement of the ship in this condition is 3870 tons; mean draught of water 19 feet 4½ inches, freeboard 2 feet 4½ inches, and the metacentric height ·85 foot.

Fig. 17 shows a transverse section of the ship in the above condition, denoting the positions of the centre of buoyancy, centre of gravity of ship and cargo, the metacentre, and the height of the respective decks.   It will be observed from Fig. 16 that the angle of maximum stability is $55\frac{1}{4}°$, and the righting moment in foot-tons, is 4218.   The angle of vanishing stability

FIG. 17.

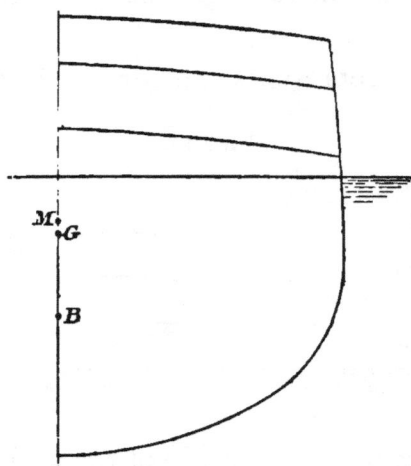

is $96\frac{1}{2}°$; and the angles at which the edges of the exposed main deck, raised quarter-deck, and bridge deck become immersed are 10°, 22°, and 30° respectively.

Curve B, Fig. 16, represents the vessel under all the conditions as described for curve A, but assuming the fore well to be filled with water amounting to 267 tons.   The displacement is 4137 tons, mean draught 20 feet $6\frac{1}{2}$ inches, the freeboard 1 foot $2\frac{1}{2}$ inches, and the metacentric height ·4 foot.

Fig. 18 is a transverse half section showing the position of the C.B. and metacentre at this draught, and also the position of the common C.G. of the ship, cargo, and water in well, and the height of the various decks at this immersion. The angle of maximum stability is reached at $46\frac{1}{2}^{c}$, and the righting moment at the angle being $1 \cdot 820$ foot-tons, the angles at

FIG. 18.   FIG. 19.

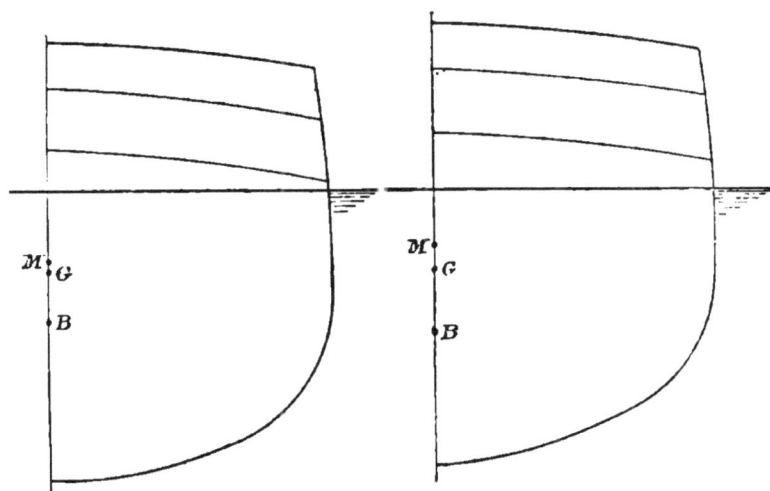

which the respective deck edges become immersed are 6°, 18°, and 27°, and the angle of vanishing stability $80\frac{1}{2}°$.

Curve C, Fig. 16, represents the vessel when laden with a heavier cargo than in curve A, the holds not being full, as in the case of the heaviest description of coals. The displacement and draught are the same as in curve A, but the metacentric height is increased to $1 \cdot 6$ feet. Fig. 19 is a transverse half-

section, showing the respective positions of the C.B.,
common C.G. of the ship and cargo, metacentre, and
decks.   The angle of maximum stability is 60°, and
the righting moment at this point is 6656 foot-tons.
The angles at which the edges of the various decks
become immersed are the same as those given in
the description of curve A ; the draught of the
vessel being the same, the range of stability being
much increased, and the vanishing point $112\frac{1}{2}$ inches.

Figs. 20 and 21 show the profile and plan of the
vessel referred to.   Taking the load displacement,
as in curves A and C, namely 3870 tons, the surplus
buoyancy due to the parts of the ship above the
water-line is 1967 tons, or 33·7 per cent. of what
would be the total displacement if it were wholly
submerged.   The portion of the ship between the
load-line and the main deck gives 560 tons displacement, or 12·6 per cent. surplus ; that due to the sheer
of the vessel is 301 tons, which added to 560 gives
18·2 per cent. ; that due to the quarter-deck is 307
tons, which added to 861 tons gives 23·2 per cent. ;
that due to the poop is 196 tons, which added to
1·168 tons gives 26·1 per cent. ; that due to the
bridge is 465 tons, which added to 1364 tons gives
32·1 per cent. ; that due to the forecastle is 138 tons,
which added to 1829 tons gives the total 33·7 per
cent. surplus buoyancy.

59.   This example is given as it represents a type of
vessel that is deservedly popular.   An inspection of
these diagrams and figures will show that the vessel

G

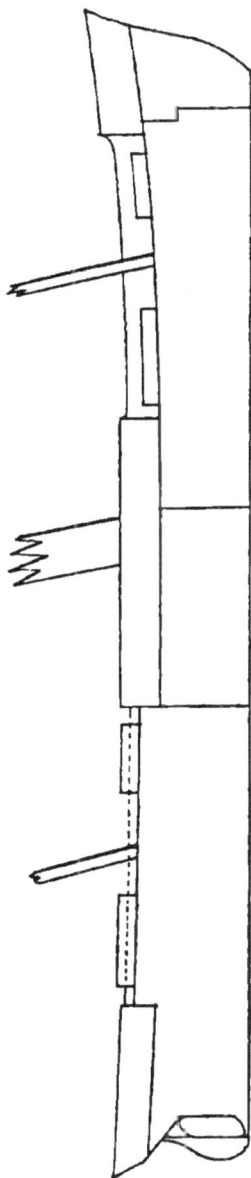

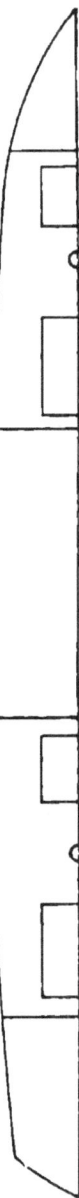

FIG. 20.

FIG. 21.

in question is a remarkably good ship; her proportions are fairly good, and in working she would give her captain little or no anxiety, providing he applied practically the lessons as taught by the curves, &c. An impression largely prevails amongst masters and officers as to the great loss of stability incurred by well-decked steamers through the fore-well filling with water. There is no doubt that filling the well with a sea is in itself a source of danger, but not by reason of the loss of stability. What happens is this: filling the well depresses the bow and raises the stern, thereby causing a loss of efficiency in the propeller and rudder; the ship loses way and steerage power, and falls off into the trough of the sea, in which condition a beam sea might strike her and cause serious damage. Many of our readers will have had experience of this.

60. The case of the well of a steamer filling with water has been exhaustively investigated by Professor Elgar, the present Director-General of Dockyards. He has considered the case of a vessel 257 feet × 35½ feet × 18½ feet with a well 60 feet in length bulwarks 5 feet high, and has assumed that there is no other outlet for the volume of water filling the well than that which it finds by pouring itself out over the bulwarks as the vessel inclines. He shows that although the volume of water which the well holds—186 tons—reduces the initial stability to nothing, and keeps the vessel unstable up to 10° of inclination, the stability becomes positive at that

angle, when only 98 tons of water remain in the well.
At 20° this water is reduced to 28 tons, and at 30°
the stability becomes the same as if the well did not
exist, and remains the same for all larger angles of
inclination.   His conclusion is that so far as stability
is concerned, the well cannot be regarded as a serious
element of danger.

61. Curve A, Fig. 22, is the curve of stability of a
cargo steamship, 289·5 feet × 341 feet × 231 feet, flush
deck type.   The position of the C.G. was ascertained

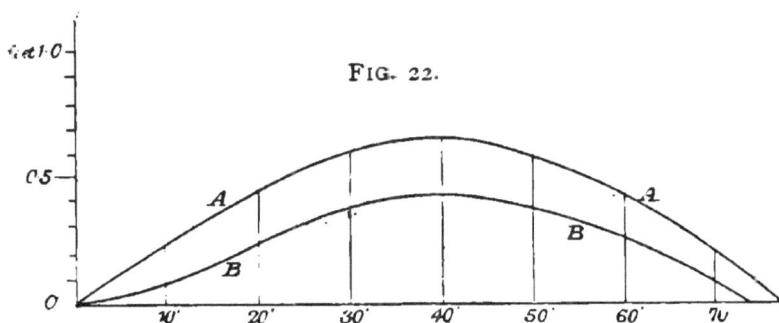

FIG. 22.

by an inclining experiment—no cargo on board, boilers
empty, but 60 tons coal in bunkers — and it was
found to be ·98 foot below the metacentre.   240
tons of coal were assumed to be placed in the
bunkers; cargo holds and tween-decks filled with an
homogeneous cargo occupying 61·2 cubic feet to
the ton.   The vessel then had a legal freeboard of
4 feet 7 inches.   It will be seen that the angle of
maximum stability is at 40°, the length of the
righting lever being ·68 foot, and the stability
vanishes at 77°.   The effect on the stability by

decreasing the breadth of the vessel by 2 feet is
well illustrated by curve B, Fig. 22, the length, depth,
freeboard, nature, and amount of cargo remaining the
same. The curve is much reduced, the maximum
length of the righting lever being but ·43 foot.

Assuming this vessel to be filled with water ballast

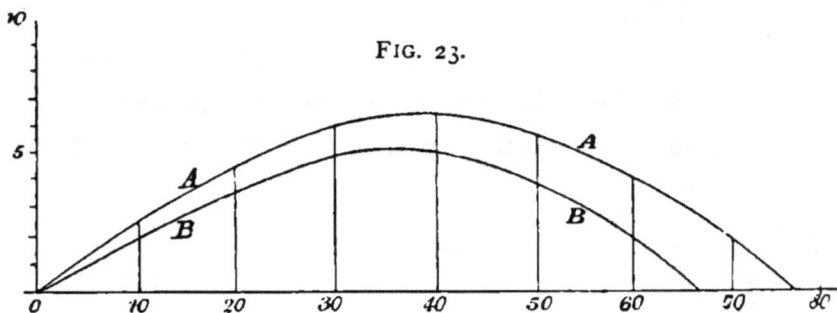

FIG. 23.

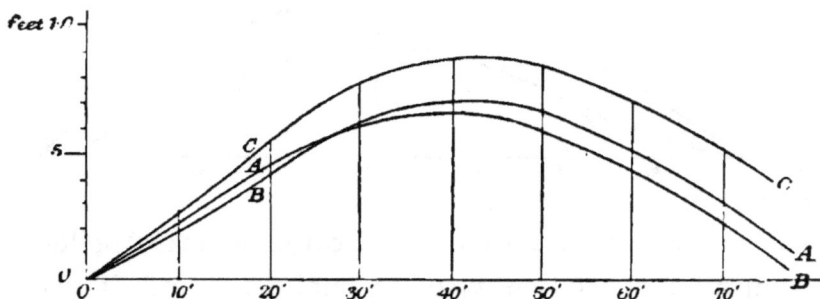

FIG. 24.

tanks 2 feet high above the floors of the fore and aft
holds, thus raising the position of the C.G. of the cargo,
the vessel's stability is reduced from curve A to curve
B (see Fig. 23). In Fig. 24 curves are shown illustrating the effect upon the stability for the same vessel,
by an increase in the freeboard, the cargo being in
each case supposed to fill the vessel; and also the

effect of variations in the density of the cargo. Curves
A and B show the difference in the stability with
6 feet freeboard, the homogeneous cargo with which
she is filled being proportionately lighter than that
which fills her with 4 feet 7 inches freeboard.
There is no material change in the stability under
these conditions, other than that a part of the area of
curve A at small angles of heel is transposed to larger
angles of heel and somewhat increases the stability at
these angles.   Curve C represents the stability when
the vessel is laden with a cargo of the same density
as in curve A, but having a freeboard of 5 feet 1 inch,
the spaces in the tween-decks at the ends of the
vessel being left empty. This curve exhibits a marked
improvement in the stability of the ship, from which it
will be seen that in cases where vessels have insufficient
stability when laden with homogeneous cargo, which
practically fills them, it will generally be effective to
restrict the amount of cargo stored in the tween-
decks.

Fig. 25 is a longitudinal section of this vessel without water ballast tanks, filled with a homogeneous
cargo which gives her 5 feet 1 inch freeboard, but
with the spaces shaded in the tween-decks left empty.
This vessel has the stability represented by curve C.
Fig. 26 shows the vessel fitted with water ballast
tanks, in fore and aft holds, which reduce the space
available for cargo, by 6000 cubic feet. Assuming
her under these circumstances to be filled with a
homogeneous cargo, which gives her 4 feet 7 inches

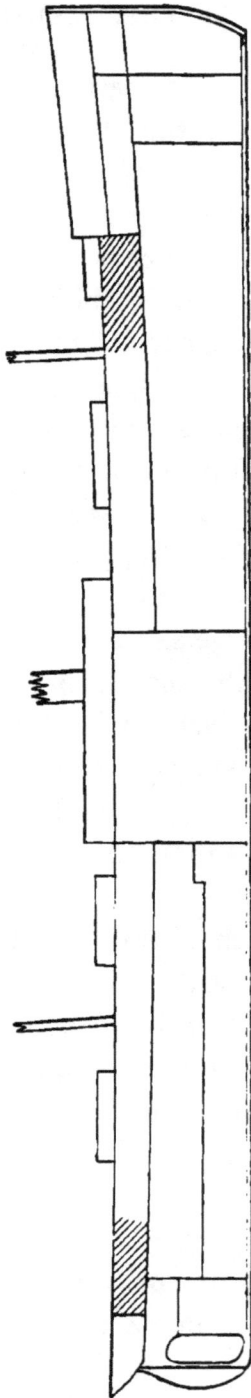

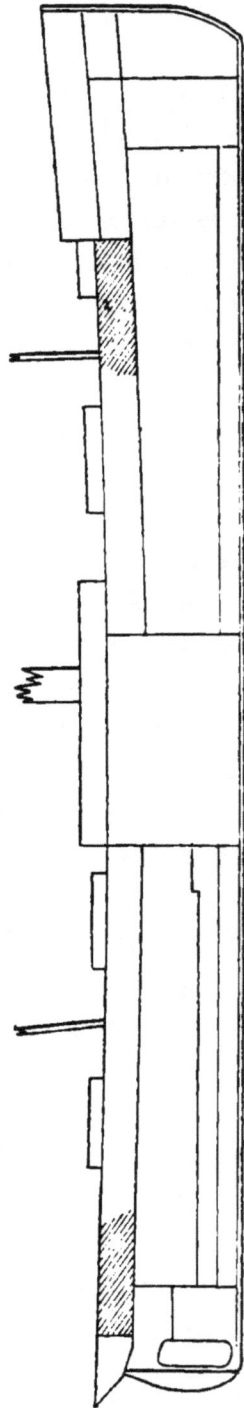

FIG. 25.

FIG. 26.

freeboard, her stability would be reduced, as shown by the curve A in Fig. 23, that is, if the spaces in the tween-decks were left empty. This shows that she could be safely laden to 4 feet 7 inches freeboard, with all cargoes which do not exceed the density of 53·4 cubic feet per ton. If 50 tons of iron ballast or an equal weight of water in the ballast tanks, be placed in the bottom of the vessel, she could load to the same freeboard with a full cargo of 61 cubic feet to the ton, and still have the same stability as is represented by this curve ; but for all lighter cargoes both the freeboard and stability would be increased.

62. The relation of beam to stability (says Sir E. J. Reed) is better shown by considering the case of a prism of rectangular section, 50½ feet beam, and immersing it 21 feet in water, leaving 6½ feet freeboard, and assuming the C.G. to be 3 feet below the water-line, and constructing for this floating body a curve of stability A, Fig. 27, it will be seen that the angle of maximum stability is 20 degrees, and the curve has a range of 38¾ feet. By increasing the beam of this floating prism by 2½ feet, the curve of stability at the same draught is represented by B in the same figures, the angle of maximum stability being the same as before, 20 feet ; but the range is extended to 41¼ feet by adding successive increments of beam of 2½ feet up to 60 feet, and retaining the same amount of freeboard. The curves C, D, E, would indicate respectively the stability due to these additions, and assuming the position of the C.G. to remain

unaltered, the position of the angle of maximum stability will remain unaltered also, although the *amount* of stability will be more than doubled, and it would absorb more than twice the amount of applied force to heel the broad prism to an angle say of 10° than it would to heel the narrow prism to the same angle. It will be observed that the curves A, B, C, D, E, Fig. 27, produced by varying the beam in this manner, rapidly leave each other at starting, converge

FIG. 27.

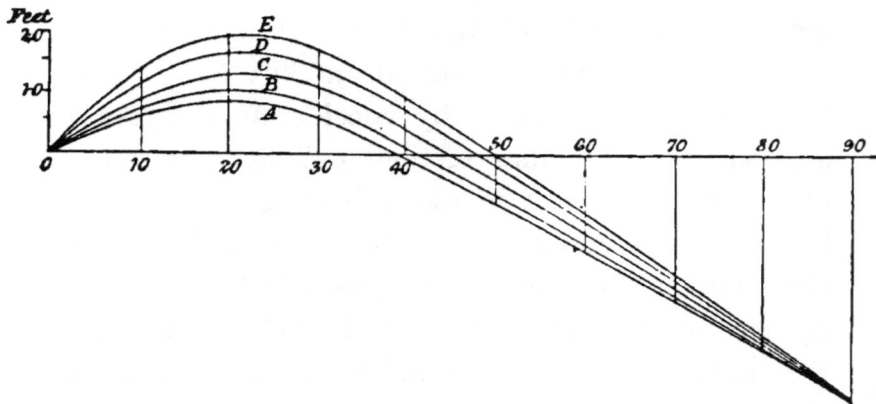

again at large angles of inclination, finally meeting at 90° or when the prism is on its side and the top and bottom become vertical, at which point the amount of instability in all cases is the same.

63. The position of the centre of gravity of the ship should be known within certain limits of the truth. To obtain the position of the C.G. is not in itself a difficult matter, and is an operation which the shipmaster of the future will perform with as little

trouble as present one "takes sights." When a vessel is being built, the position of the C.G. is found by calculation, but such is the increasing size and the enormous variety of fittings introduced, the weight of each individual article in the hull and equipment having to be known, that the process is tedious and the margin of error large.

The C.G., we have said, is a point at which the whole weight of the ship may be considered to act. The definition is not the best, but will suffice for our present purpose. The position of the C.G. is calculated with reference to two planes: the horizontal plane at the top of the keel, and the vertical midship longitudinal plane. It is obvious that if the ship is symmetrically built (all ships are not), and the machinery is equally distributed on both sides; the C.G. will lie somewhere in the vertical longitudinal midship plane. In any well built and designed vessel we may assume that the C.G. lies somewhere in this plane. We then have to locate the position of the C.G. with reference to the horizontal plane. The intersection of these two ordinates will be the position of the C.G. In practice this point is found by merely shifting a known weight and measuring an angle, and the principle is the same as that illustrated on pages 50–1, in which we showed that, in any given mass, the transfer of a portion to another position was accompanied by a shifting of the C.G. by an amount depending upon the respective weights and the distance between the position C.G. of the original

and transferred portions. Applying this to an actual
ship, let G, Fig. 28, be the position of the C.G. of the
ship and of all weights on board, W L the water-line,
B the centre of buoyancy.

w is a weight placed on the upper deck, or tween-
decks ; the weight is of course known accurately, and
its position, or rather the distance of its C.G. from the

FIG. 28.

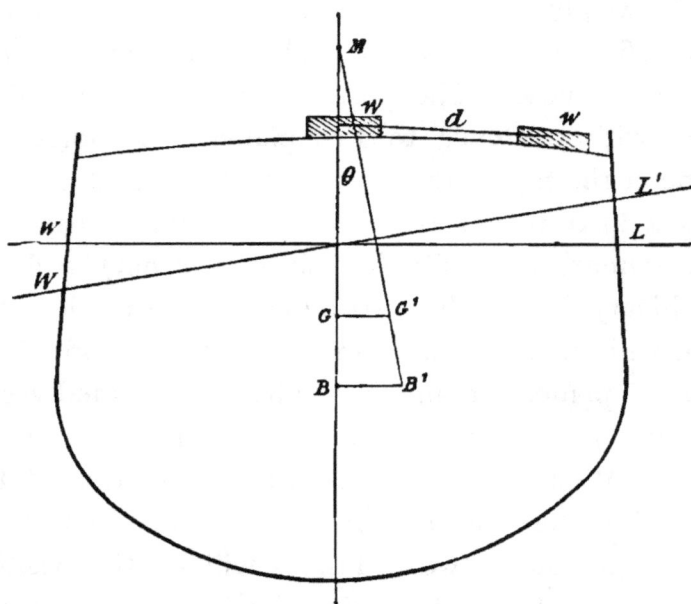

midship line, is likewise known. If w be shifted
from its position to one side, it will cause the vessel
to take a list. This angle of heel can be accurately
measured by means of a plumb bob ; or a good
observer with the sextant can station himself in a
favourable position and, with the sextant firmly fixed,
measure the angle between the ship's rail and some

distant vertical object when the ship is upright, and
then, when the inclination is effected, the difference
between the readings will give the value of the angle.
Let the angle of inclination be θ, let $W_1 L_1$ be the
new water line, and G_1 the position of the new centre
of gravity, and D the total displacement in tons. We
have then

$$D \times G\,G_1 = w \times d.$$

but $G\,G_1 = G\,M \tan \theta,$

$$\therefore D \times G\,M \tan \theta = w\,d$$

and

$$G\,M = \frac{w\,d}{D \tan \theta}$$

The position of B is calculated from the drawings
of the ship. Suppose we have a vessel whose displacement is 3000 tons, and she has the height of the
metacentre above the centre of buoyancy of 14 feet,
by shifting a weight composed of 15 tons of fire-bars
through a distance of 20 feet we incline the ship $7\frac{1}{2}°$.
We have

$$G\,M = \frac{w\,d}{D \tan \theta}$$

$\log 15 = 1 \cdot 17609$	$\log 3000 = 3 \cdot 47712$
$\log 30 = 1 \cdot 30103$	$\tan \theta = 9 \cdot 11942$
$2 \cdot 47712$	$2 \cdot 59654$
$2 \cdot 59654$	

$$9 \cdot 88059 = \cdot 758 \text{ foot}$$

$$\therefore \quad G\,M = \quad \cdot 758 \text{ foot}$$

$$\text{and} \quad B\,G = 13 \cdot 242 \text{ feet}$$

Knowing the height of B (the centre of buoyancy) above the keel, and the distance B M, the position of G, the centre of gravity of the ship, is found as above. It is sometimes necessary to know how much inclination will be produced on shifting a certain weight through a given distance. In oil steamers it might be necessary to know whether, under given conditions, a side tank could be filled. If we know approximately the metacentric height, the position of the centre of gravity, and the displacement in tons, we can easily find out the angle a given weight will incline the ship to.

Example : To effect some repairs to a coal bunker it is found necessary to shift about 20 tons of coal in the tween decks in a horizontal direction through a distance of 15 feet. The displacement of the ship is 850 tons, the metacentric height is 9 inches (= ·75 foot). What will be the inclination produced ?

Referring to the formula

$$G M = \frac{w \times d}{D \tan \theta}$$

$$\theta = \tan^{-1}\left(\frac{w \times d}{G M \times D}\right)$$

$$\therefore \tan \theta = \frac{w \times d}{G M \times D}$$

$$= \frac{20 \times 15.}{·75 \times 850} = \frac{300}{637·5}$$

$$\log 300 = 2·4771$$
$$\log 637·5 = 2·8044$$

$$\overline{\qquad\qquad}$$

$$9·6727$$

Angle of keel = 25° 12'

64. Thus far we have dealt only with the statical stability of a vessel, and have defined the points M, G, and B. With reference to any longitudinal inclination, it will not be difficult to see that precisely the same rules apply, when making a vessel to incline by the head or stern, and indeed to discuss the subject fully we should have to discuss the subject of longitudinal stability.   Our purpose, however, has been merely to give the simplest view of the subject for the purpose of applying the doctrine of stability to oil-carrying steamships.   We have not touched at all upon the subject of dynamical stability.   Those of our readers who wish to read up this most interesting and important question of stability of ships, should obtain the works of Reid, White, Thearle, &c.

65. When the skin plating of a ship is pierced, through collision or stranding, the water will enter the ship with a velocity and amount depending on the area of the hole, and the depth below the water-line. Of course, when damage is done, we cannot waste time in making measurements ; but it is always well to be acquainted with what is taking place, and so we give the following rule.   To determine the quantity of water that will enter the ship at a given depth in a certain time, let

A = area of the hole in square feet (estimated),
d = depth of centre of hole below water-line,
v = velocity of water in feet per second,
$v^2 = 64d$ (approximately),
$\therefore\ v = 8\ \sqrt{d}.$

So, when a hole is made in a vessel's plating, the volume of water entering per second = 8 $\sqrt{d}$ A.

Example: In catting the anchor, the fall carries away, and in dropping the anchor-fluke pierces the ship's side near the 10 feet water-line mark. The hole is triangular, is apparently 6 inches each way; ship draws 21 feet. We may assume the area of the hole, = 18 square inches = ·125 square feet.

Water entering per second,

= 8$\sqrt{d}$ A.
= 8$\sqrt{11}$ × ·125
= 8 × 3·31 × ·125 = 3·31 cubic feet per second;

or 198·6 cubic feet per minute, = 56·7 tons per minute, would be the amount of water entering the fore peak.

A 7-inch Dounton, making thirty revolutions per minute, will lift forty-five gallons per minute, so, to keep such a leak under, we should require about thirty pumps of this pattern. Whereas a 10-inch Worthington would keep such a leak under with ease.

CHAPTER III.

66. WE are now in a position to investigate the conditions of stability of petroleum-carrying steamers, so far at any rate as will ensure safe handling when loading and discharging. It is without the scope of this work to discuss the forces that vessels are subject to when in heavy weather : in point of fact, very little is known at all of the forces acting on the hull of a vessel when in a sea-way. We shall therefore confine ourselves to known or approximately known quantities. The subject of the stability of oil-steamers has been exhaustively treated by Professor Jenkins and Mr. Martell, by whose permission the author reproduces parts of their well-known papers.

67. It will be well to discuss some of the questions

relating to ballast tanks, inasmuch as many petroleum vessels are constructed with the doubtful advantage of having double bottoms. We say doubtful advantage, as while the fact of a ship having a double bottom adds enormously to her structural strength, yet, in the case of a petroleum-carrying steamer, in the event of the oil finding its way into an empty ballast tank, we have a most dangerous state of things occurring : we have, in effect, two compartments, the ballast tank and the oil tank, partially filled. It will be seen what effect this has upon the stability of the ship.

68. There are three cases to be considered in relation to compartments. A compartment may be completely filled with oil or water ; it may be partially filled, but no communication with the outside water ; and lastly, a compartment may have the water in full communication with the outside water. The first case is that of ordinary ships' tanks. The effect on the stability of a ship by filling the ballast tank is the same as would be produced if any other homogeneous body of the same weight as water were used. In the second case, we have a tank partially filled, the enclosed oil or water altering its position with every change in the position of the ship. This occurs in practice when filling or emptying ballast and oil tanks. Taking the case of a ballast tank, let W L, Fig. 29, be the water-line when the ship is upright, wl the surface of the water in the partially-filled ballast tank, G and B the centres of gravity and buoyancy when the ship is upright, and M the corresponding meta-

H

centre.   Suppose now, in the course of filling the tank
(the ship being light), she heels over through a small
angle, as is generally the case, and let W' L' be the
new water-line, and $w' l$ the surface-line of the water
in the ballast tank, and B' the new centre of buoyancy.
As is easily seen from our previous remarks, the C.G.
of the ship will no longer be at G, but at G', along

FIG. 29.

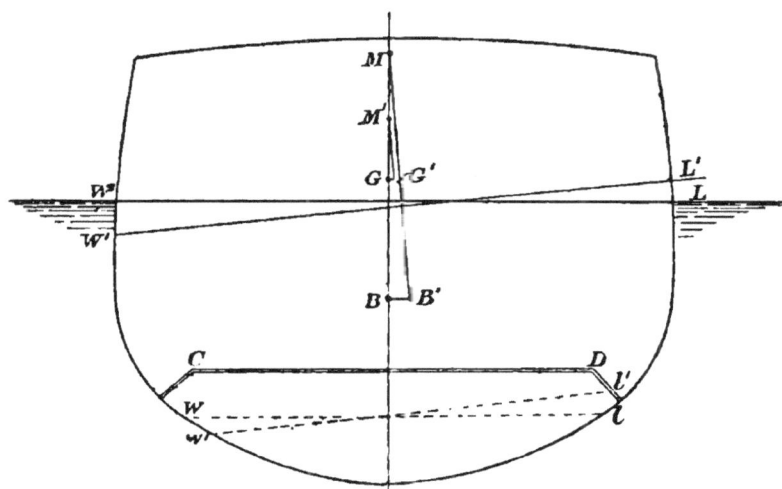

a line nearly parallel to B B'.   Let M' be the
intersection of the vertical through the point G'.
The angle being small, $BM = \frac{2}{3} \int \frac{y^3\, d\,x}{D}$.   Where
y is the half-breadth at the water-line, and D
is the displacement.*   In the present case G M' =

---

* The symbol $\int$ is the sign of integration, and is used in the integral
calculus.   It means that a number of small quantities are to be added
together.   By the integral calculus we ascertain from the ratio of
indefinitely small changes in two or more magnitudes the function (f)
which governs the changes.

$\frac{2}{3} \int \frac{y_1{}^3 \, dx}{D}$, where y_1 is the half-breadth of the free water surface, and when G M′ is less than G M (or when $\frac{2}{3} \int \frac{y_1{}^3 \, dx}{D}$ is less than $\frac{2}{3} \int \frac{y^3 \, dx}{D} - $ B G), the ship will float safely. If G M′ is greater than G M, the ship is unstable; and when $G M_1$ is equal to G M, the equilibrium is indifferent.

69. Without going into the mathematics of the question, and remembering what we have previously said, that the point M denotes the limit of the range of the C.G., it will easily be seen from an inspection of the Fig. how attended with risk it is to have a large ballast tank partially filled. Although the vessel is only inclined through a small angle, yet M drops to M′; in fact, when filling ballast tanks, they should always be filled one by one, and the vessel be kept on an even keel, as otherwise the metacentre falls rapidly. It must be borne in mind, too, that we, in an elementary sketch like the present, only treat this subject from the *statical* point of view : that is, we assume we have our vessel in smooth water, and that there are no other disturbing elements present, and that we are dealing with known forces and weights ; to take into account what forces are acting upon a ship in a sea-way, would lead us on to the path of dynamical stability. We see how rapidly the metacentre falls when filling a tank under favourable conditions. The number of vessels laden with homogeneous cargoes that have foundered at sea through the loss of

stability occasioned in the very attempt to obtain it, is
beyond comprehension.   Captains should always bear
in mind that the mere fact of admitting water or oil
into an empty tank causes a loss of stability till the
tank is full.   It will be seen further on that, according to the investigations of Professor Jenkins and
Mr. Martell, the question of loading and discharging tank steamers requires the most careful
attention.

70. As improvements in the design of cargo steamers
are being gradually introduced, it is certain that
before long the principle of making the various decks
perfect water-tight flats will be largely utilised.   It is
done at present, to a limited extent, in the latest
types of fast mail steamers, and we may soon expect
to see the principle extended to cargo steamers.   As
regards oil steamers the tendency is to increase the
size, and, in the writer's opinion, we shall have to
introduce horizontal oil-tight flats, in addition to
longitudinal and transverse bulkheads.   It may therefore be interesting and useful to investigate, or rather
to give Sir E. J. Reed's investigation of the case in
which a vessel has a water-tight flat extending across
the ship, similar to that which exists in many war-
ships, and situated a short distance below the load
water-line.   If we imagine an ordinary cargo boat has
the hatches in her iron tween-decks perfectly water-
tight, and that from some cause, such as shipping a
sea, a large volume of water has found its way into
the tween-decks, which we will regard as free from

cargo and coals, let W L be the water-line, Fig. 30, *w l* the surface of the free water in the tween-decks, B and G are the centres of buoyancy and gravity, and M the corresponding metacentre. The vessel takes a list, and the corresponding C.B., is now at B'; G will shift to G', along a line parallel to B' B.

It is easily seen that the mass of water in this case

FIG. 30.

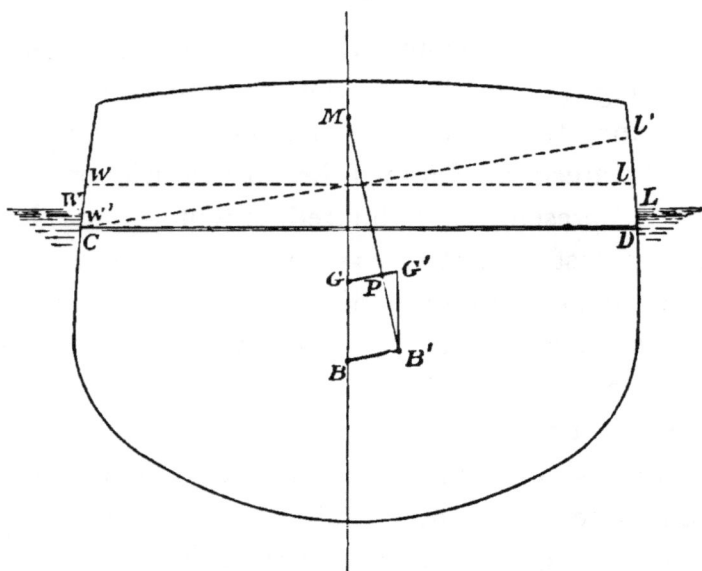

is upsetting the ship, and the moment of the upsetting couple is G' P × ship's displacement. Calling the angle of heel θ, and D the displacement of the ship, we have the moment of the upsetting couple

$$= G' P \times D = D \times B' G' \sin \theta$$
$$= D \times B G \sin \theta.$$

It will be gathered from what we have said that in

a petroleum bulk steamer it is not advisible to have
large volumes of liquid free to take any position due
to rolling of the ship.   Safety in handling is best
assured by a good system of subdivision.

71. Remarks by Professor Jenkins.—Referring first
to the stresses produced in bulkheads, he says :—

"During the process of filling or discharging, a
bulkhead can never have to withstand the pressure of
a greater head of oil or water than that corresponding
with the level of the oil or water in the reserve tubes.
At sea, so long as adjacent compartments are full, the
dividing bulkhead can have very little stress to bear,
as it is pressed about equally on both sides.   The
bulkheads which bound the oil hold have the oil only
on one side of them, and are subject to considerable
stress, and more especially the aftermost on account
of its greater breadth, during the whole of the voyage
when in the laden condition.   Also the bulkheads
which bound the water ballast tanks when these are
filled and the ship light, are similarly circumstanced.

"But the pressure due to the head of oil or
water, as the case may be, does not represent the
whole pressure which a bulkhead has to bear,
and it may be of interest to refer briefly to them
here.   Let us suppose a vessel laden and about to
start on her voyage.   As she acquires onward motion
each element of cargo has to acquire an equal velocity
in the direction of motion.   Ordinarily the force
causing the onward motion is communicated to the
vessel herself from the shaft through the thrust block

to the hull, and from the ceiling and decks to the cargo by virtue of frictional resistance. But with liquid cargoes such cannot be the case. This will be well understood if we imagine a vessel containing free water to be set in motion. The water will not at first partake of the motion of the ship, but will move aft, heaping itself up against the aftermost bulkhead of the compartment, presenting to the eye of an observer an inclined surface. In such a case the bulkhead has to supply the necessary accelerating force to the water, and in doing so it is itself strained. Similarly in an oil vessel, owing to the fluidity of the cargo, nearly the whole of the force necessary to cause the oil to acquire onward velocity at the same rate as the ship herself has to be communicated through the medium of the bulkheads. As the vessel begins to acquire motion, the oil in each compartment lags behind, it presses against the bulkhead, the bulkhead is to some extent deflected, and when the increase of stress in the bulkhead is just that caused by a pressure great enough to communicate the necessary increment of velocity to the contained oil, it yields no more. It might at first sight be thought that the pressure on the bulkheads from this cause as we move aft will be intensified—that the pressure on the bulkhead on the after side of the foremost compartment, for instance, in causing it to deflect, would be communicated to the next compartment, and so on, and that there would thus be an accumulation of pressure on the aftermost bulkhead. No doubt if the tanks were filled and sealed some such action as I have sketched would take place ; but since the oil surface in each tank is free, the bulkhead is free to deflect, and any such deflection is measured by a slight rise in the level of the oil in the reserve tube (expansion chamber). Practically, therefore, the increase of pressure on the aftermost bulkhead of each compartment from this cause is due almost entirely to the oil of that compartment. The effect of moving ahead is thus to increase the stress in the divisional bulkheads, although the increment is usually not great in amount. The aftermost bulkhead is in this way subject to a pressure greater than that due to the head of oil in the reserve tube, and the foremost bulkhead to a pressure somewhat less. A similar kind of action goes on when the vessel is being brought to rest, except that then the bulkheads separating the oil compartments are strained in the contrary way, and the actual stress on the foremost bulkhead is increased, that on the aftermost being diminished.

"Again, if we suppose an oil vessel to be rolling in the trough of the sea, and to partake of the orbital motion which the sea water has, the effect of this motion is to increase the apparent weight of the contained oil when the vessel is in the trough of the wave by as much in some cases as 20 per cent., and to diminish it to a corresponding extent when the vessel is on the crest. Practically, the effect of the alternate increase and decrease in the apparent weight affects only the bulkheads bounding the oil hold, increasing the pressure in the wave trough nearly into equality with that due to an equal head of water, and similarly decreasing it on the wave crest. In the same way on the return journey the bulkheads bounding the water ballast may be subject to stress due to the pressure of a liquid of proportionately greater density than water.

" But although the stresses induced by these causes are important in themselves and deserve consideration, it is not for the purpose of withstanding these alone that bulkheads of oil-carrying steamers need the strengthening usually applied to them. A more serious condition of affairs would be reached if by any accident a compartment became only partially filled at sea, owing to the longitudinal motion of the contained fluid caused by the pitching and ascending of the vessel in crossing waves, and it is mainly to provide against this contingency that the elaborate arrangements to be seen in most oil-steamers are supplied. In any case, however, the bulkheads which should receive most attention and be most strengthened are the two which bound the cargo hold, as well as those which bound that part of the hold appropriated for ballast on the return journey. Whatever may be the condition of affairs, the intermediate bulkheads can never possibly be subject to straining action to the same extent as the ones in question."

72. So far as those vessels which carry water ballast in tanks below the oil compartments are concerned, it may be noted that the tops of the tanks have to bear

continuously the stress due to the head of oil resting upon them, and extending in these instances to some distance above the upper deck.   In the trough of the wave, too, the tank top has to sustain the augmented pressure due to that increase attained in which the centre of gravity and centre of buoyancy are again in the same vertical line.   In the case, however, in which

FIG. 31.

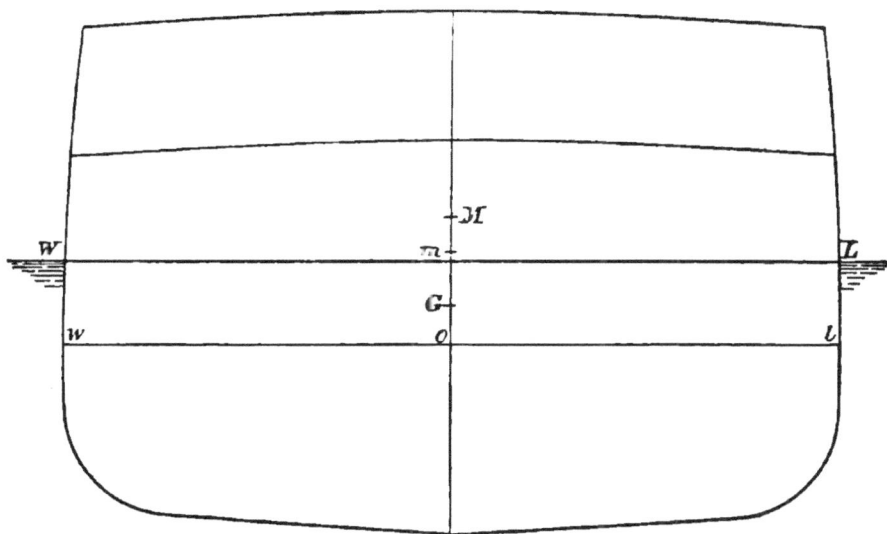

a vessel contains water or oil having a free surface, she may not remain upright even though the common centre of gravity be below the metacentre, and this is due to the fact that the contained liquid is constantly ready to change its surface in order that it may remain horizontal, or parallel to that of the water in which the vessel floats.   Take for example the vessel Fig. 31, M being the metacentre and G the common centre of gravity of hull and cargo when the vessel is upright. Let W L be the water-line, and $w\,l$ the level of the liquid cargo in one of the compartments, If the vessel be forcibly inclined from the upright position through a small angle, so that $W_1\,L_1$, Fig. 32, becomes the water-line, the free liquid surface will follow it and assume the parallel position $w_1\,l_1$. This

FIG. 32.

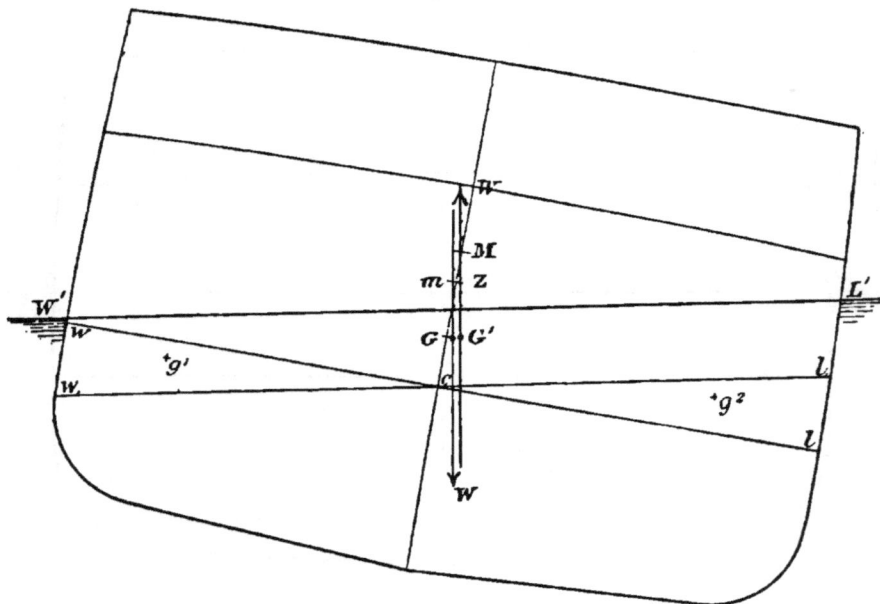

cannot be accomplished without some movement in the liquid, the quantity of oil on the emerged side being diminished by the wedge $w\,o\,w_1$ whose centre of gravity is at g_1, and that on the immersed side being immersed by an equal wedge $l\,o\,l_1$ whose centre of gravity is at g_2. Hence the common centre of gravity of hull and cargo must move from G to G_1, the line $G G_1$ being parallel to $g_1 g_2$, and its length being to $g_1 g_2$ as the weight of the wedge of oil shifted is to the weight of the laden vessel. The vertical through G_1 will cut the middle line at some point m, and through this point the weight of the ship and cargo acts at the given inclination. The point m is related to G in the same way as the metacentre to the centre of buoyancy, that is to say, whatever be the inclination of the vessel, so long as it be small, the vertical through the new centre of gravity will always cut the middle line at the same point m. The resultant upward pressure of the water will of course act through M, and hence we see that if m be below M the vessel will return if released, to the upright position, the wedge, $l o l_1$, of contained liquid passing to its former position, $w o w_1$; if m coincides with M the vessel will not so return, but will remain in neutral equilibrium ; and if m be above M the vessel will incline still further from the upright position, coming to rest only when the centre of buoyancy is drawn sufficiently far on the immersed side to be in the vertical line through the centre of gravity of hull and cargo. We see then that the point m has the same property in determining the character of the equilibrium of a vessel containing liquid having a free surface, as the centre of gravity has in the case of a vessel containing cargo which does not shift. This interval m M, the learned professor gives the name "effective metacentric height," inasmuch as this distance affords a measure of the vessel's initial stiffness, in the same way as the metacentric height does under ordinary circumstances. The righting moment at a small inclination θ from the upright is in fact W m M sin θ. It is not difficult to measure accurately the distance G m, which, as remarked before, is constant for all small inclinations of the vessel with a given extent of free surface.

73. Let us assume that the compartment containing the free liquid is rectangular in form at the height of the oil surface, its length being l and half breadth y. Let W be the displacement of the vessel in tons. Suppose the vessel inclined from the upright through a very small angle θ, Fig. 32. The volume of the wedge of oil, $w\,o\,w_1$, transferred to $l\,o\,l_1$, is $\frac{1}{2}\,y^2\,l\,\theta$, and the distance through which its centre of gravity is ultimately transferred is $\frac{4}{3}\,y$. If the contained liquid were salt water, the weight of the wedge would be $\frac{\frac{1}{2}\,y^2\,l\,\theta}{35}$, and its moment $\frac{\frac{1}{2}\,y^2\,l\,\theta}{35} \cdot \frac{4}{3}\,y = \frac{\frac{2}{3}\,y^3\,l\,\theta}{35}$, and, from what has gone before, this product must be equal to W . G G_1. But since the density of petroleum is only about $\cdot 8$ that of salt water, the weight of the transferred wedge will be but

$$\frac{\frac{1}{2}\,y^2\,l\,\theta}{35} \cdot \frac{8}{10},$$

and the moment

$$\frac{\frac{1}{2}\,y^2\,l\,\theta}{35} \cdot \frac{8}{10} \cdot \frac{4\,y}{3} = \frac{\cdot\,y^3\,l\,\theta}{35} \cdot \frac{8}{10}$$

Equating

$$\text{W}.\text{G }G_1 = \frac{\frac{2}{3}\,y^3\,l\,\theta}{35} \cdot \frac{8}{10};$$

and $G G_1 = G m \cdot \theta$ ultimately.   Hence,

$$G m = \frac{\frac{2}{3} y^3 l}{35 \text{ W}} \cdot \frac{8}{10} \cdot$$

Now $\frac{2}{3} y^3 l$ is the moment of inertia of the plane of the free oil about the middle line ; call it I_1, and we may then write :

$$G m = \frac{8}{10} \cdot \frac{I_1}{35 \text{ W}} \tag{1}$$

As an example, take a vessel 40 feet broad, having a compartment 24 feet long amidship, in which oil is free to move from side to side, the displacement being 3000 tons.   Here $I_1 = \frac{2}{3} (20)^3 24$, and by substituting in equation (1), we find $G m = \cdot976$ feet.   Hence, in such a case, if the height of the metacentre above the centre of gravity be less than $\cdot976$ feet, the vessel will not remain upright under the above conditions. If there be two such compartments of equal length partially filled, the height of m above G will be doubled, and the actual metacentre height must exceed $1 \cdot952$ feet if the vessel is to remain upright. Similarly as regards a greater number of compart- ments.   It may be added that, although equation (1) has been established only for the case in which the free surface of the oil is in form a rectangle, the equation is rigorously true whatever be the form, I_1 being then the moment of inertia of the plane of the free oil surface above the middle line.

The above investigation is for the case in which the

vessel has no middle line bulkhead, or a bulkhead
perforated so as to admit of the passage of oil readily
from one side to the other.   Let us now examine the
case in which the compartments are divided by an oil-
tight or practically oil-tight middle line bulkhead.   In
that case, if the vessel of Fig. 31 be inclined through

FIG. 33.

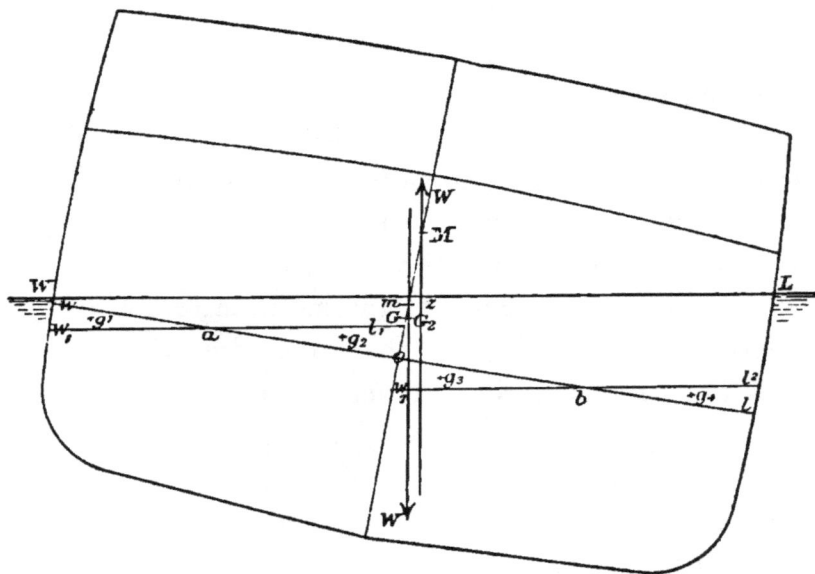

a small angle, the oil on the one side of the middle
line bulkhead will change its surface from $w\,o$ to $w_1\,l_1$
parallel to $W_1\,L_1$, and that on the other side from $o\,l$
to $w_2\,l_2$, also parallel to $W_1\,L_1$, Fig. 33.   Here we
have one wedge $w\,a\,w_1$ shifted to $o\,a\,l_1$, and another
wedge $w_2\,o\,b$ moved to $l\,b\,l_2$, and their combined effect
is as before to make the centre of gravity G move

away from its position in the middle line to some point G_2. The distance $G G_2$ is not, however, so great as $G G_1$ in the preceding instance. The wedge of oil shifted has only one-fourth the volume of the wedge of the previous example, and the distance through which its centre of gravity is moved is only one-half as great. On the other hand, two wedges are shifted here instead of one. Hence we shall obtain by the same reasoning as before

$$G G_2 = \frac{\frac{1}{2} \left(\frac{y}{2}\right)^2 \theta \cdot \frac{2}{3} y \cdot 2 \cdot \frac{8}{10}}{35\ W}$$

and

$$G m = \frac{1}{4} \cdot \frac{I_1}{35\ W} \cdot \frac{8}{10},$$

or the loss of effective metacentric height is only one-fourth as great as when the oil is free to shift from side to side of the vessel. It follows that in the numerical example already quoted such loss would amount to only $\cdot 244$ feet instead of $\cdot 976$ feet. Put in another form, the loss of effective metacentric height due to the readiness of the oil to shift in a vessel with four compartments only partially filled with oil, but having an oil-tight middle line bulkhead, is the same as in a vessel of the same displacement having only one such compartment filled, but having no middle line bulkhead, or one through which the oil may readily pass from side to side.

We will only briefly advert to the case in which the oil does not extend to the bottom plating, but is

separated from it by a cellular bottom, as in some
vessels built in Sweden and on the east coast, or by
a tank having a section like an inverted V, as in a
number of vessels built specially for this trade.   Here
if the oil leaks through the flat on which it rests into
the space below in any quantity, it will be free to
move from side to side as well as the oil in the compartment above.   The effect will be to further reduce
the effective metacentric height.   On the other hand,
the transference of some oil from the free surface in
the compartment to the cellular bottom has the effect
of lowering the centre of gravity, and thus adding
to the vessel's stiffness.   If there be no oil-tight
middle-line bulkhead, the loss of effective metacentric
height is

$$\frac{I_1 + I_2}{35\ W} \cdot \frac{8}{10} - \frac{w\,x}{W},$$

where I_1 has the value as before ; I_2 is the moment of
inertia of the plane of the oil surface in the cellular
bottom above the middle line, with the weight of the
oil collected in the cellular bottom ; and x the distance
between its centre of gravity and that of the same
quantity of oil lost in the compartment above.

If the middle line bulkhead be oil-tight under the
above condition, the oil in the cellular bottom being
free to move from side to side, the loss of effective
metacentric height is

$$\frac{1}{4}\,\frac{8}{10}\,\frac{I_1}{35\ W} + \frac{8}{10}\,\frac{I_2}{35\ W} - \frac{w\,x}{W}.$$

In this case, the vessel is permanently inclined to, owing to the movement transversely of the centre of gravity of the oil which has passed into the cellular bottom.

74. We have seen that if G m be greater than G M, the vessel will not remain upright but will incline away from that position, reaching a position of equilibrium at a greater or less inclination, according to the circumstances of the case ; and it will perhaps appear obvious to those who have followed us thus far, that it is of very great import in loading or unloading, if the greatest number of tanks are to be filled or discharged simultaneously, that care be taken to maintain the common centre of gravity of hull, of coal, and fittings in the middle-line plane. If this be not done the vessel will incline sometimes through a large angle even though m be below M. To show that such is the case let us suppose the vessel to be upright and at rest, and to have a definite effective metacentric height M m, one or more of the tanks being only partially filled. Let now a small weight, as for instance some bunker coal, be moved from the middle line to one side, or from one side to the other. We should find in the case of a vessel having cargo that could not shift that she would assume a position of equilibrium at a definite inclination as in the inclining experiment. But in an oil steamer having tanks only partially filled, the inclination of the vessel due to the moving of the weight causes a wedge of oil to pass from one side to the other. This has the effect

of further inclining her—more oil consequently passes from one side to the other; she becomes still more inclined—this leads to the transference of still more oil; and so on, until finally a position of rest is attained. It is not difficult, in cases where the ultimate inclination is not great, to determine the precise inclination at which the vessel is brought to rest under such circumstances.

Suppose the vessel to be at first upright and at rest, and let a small weight, w, be moved transversely through a distance, a. This will cause the common centre of gravity of hull and cargo to move transversely through a distance $\dfrac{w\,a}{W}$. Let θ be the inclination at which the vessel is finally brought to rest. The weight of the wedge of oil transferred from the one side to the other is $\dfrac{\frac{1}{2}y^2\theta\cdot l}{35}\cdot\dfrac{8}{10}$, and the shift of the common centre of gravity due to this cause is $\dfrac{8}{10}\cdot\dfrac{I_1\cdot\theta}{35W}$, as before.

Hence the total shift of the centre of gravity is

$$\frac{w\,a}{W} + \frac{8}{10}\frac{I_1\cdot\theta}{35}\frac{}{W},$$

But since the vessel is brought to rest at the inclination θ, which is assumed not to be great, the centre of gravity must be finally in the vertical through the metacentre, and hence the shift of G may be written as $G\,M\cdot\theta$.

Equating these two values, we get

$$G M \cdot \theta = \frac{w \, a}{W} + \frac{8}{10} \frac{I_1 \cdot \theta}{35 \, W},$$

from which

$$\theta = \frac{35 \, w \, a}{35 \, W \cdot G M - \frac{8}{10} I_1}.$$

As an example, suppose the displacement of the vessel to be 3000 tons, $G = M \, 1 \cdot 25$ feet, and the compartment 24 feet long as before in a vessel 40 feet broad. Let the small weight, w, be 5 tons, and the distance through which it is moved transversely 30 feet. Ordinarily the transference of this weight of 5 tons would incline the vessel through about $2\frac{1}{4}°$, whereas from the above equation it is found that with free oil in one compartment the inclination is raised to as much as $10\frac{1}{2}°$.

For the similar case in which the same compartment is divided by an oil-tight middle-line bulkhead the corresponding equation giving the inclination will be

$$G M \cdot \theta = \frac{w' \, a}{W} + \frac{1}{4} \frac{8}{10} \frac{I_1}{35} \frac{\theta}{W},$$

from which

$$= \frac{140 \, w \, a}{140 \, W \cdot G M - \frac{8}{10} I_1},$$

If this equation be applied to the above example, it is found that the ultimate inclination of the vessel having an oil-tight middle-line bulkhead is about $3°$.

The above formulæ give the inclinations at which the vessel comes to rest under the several conditions with great closeness in all cases in which the inclination is not large, and especially when the free surface of the oil is not very near the crown of the tank. In those cases in which the free surface of the oil is very near the top of the tank, or where the inclination becomes great, its actual value can only be determined by the calculation of the vessel's stability as affected by the movement of the oil, the inclination at which the curve of stability cuts the base line being that at which the vessel comes to rest.

The above examples show the great importance of providing in the design of an oil vessel that the common centre of gravity of the hull, coal supply, stores, &c., shall be in the middle-line plane. Further, before loading or unloading cargo, if the ability of the vessel to load or discharge simultaneously from the greatest number of tanks without danger or inconvenient inclination is to be regarded, care should be taken by properly trimming the coal to bring the vessel upright before loading or unloading is begun.

75. We have dealt up to the present with the initial stability of vessels containing oil having a free surface, or with moderate inclinations only from the upright position ; it will be proper now to consider its inclining effect at larger inclinations. Although so long as G m is less than G M the vessel will continue to remain upright, she is not so capable of resisting

inclination due to external forces—such as wind pressure, or the heave of the sea—as if the cargo were not free to shift. In the latter case for moderate inclinations the righting moment may be expressed by the product of $W \cdot GM \sin \theta$, whereas in the former it is but $W \cdot m M \sin \theta$, as has been already shown. For large inclinations the above expressions cease to be correct, and the loss of righting arm caused by the shifting of the cargo can then be best represented by means of a diagram. Figs. 34 to 37, represent the stability of a petroleum vessel under a variety of circumstances. She is supposed to have arrived at New York with enough coal in the lower bunkers for the return journey to Europe. Her curve of righting arm in that condition is A, Fig. 34. The filling of one compartment 24 feet long is then begun. When the oil has risen to a height equal to one-fourth the depth of the tank, the curve of righting arm, assuming the cargo not to shift, is shown by A, Fig. 35. With the cargo free to shift, and with an oil-tight middle-line bulkhead, the curve falls to B, Fig. 35, and with the middle-line bulkhead perforated it is further reduced in height to C, Fig. 35. When the oil has risen half way up the tank the curves of righting arms under the above three conditions are A, B, and C, Fig. 36. With the oil filling three-fourths of the tank the curves are A, B, and C, Fig. 37; and with the tank full, so that shifting cannot take place, the three curves come together, forming B, Fig. 34.

FIGS. 34.

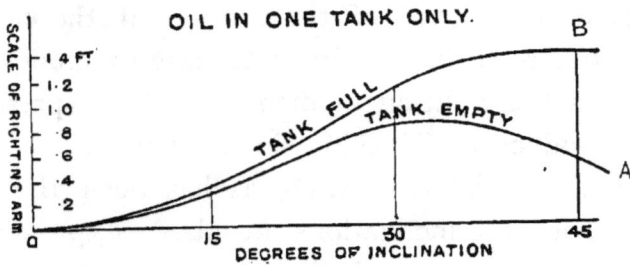

FIG. 35.

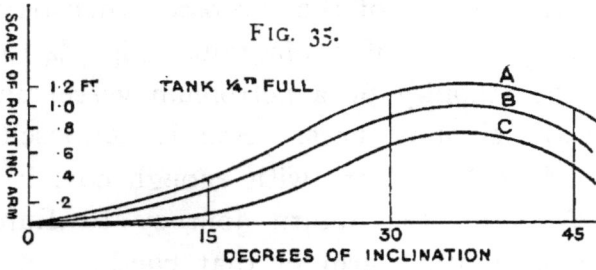

FIG. 36.

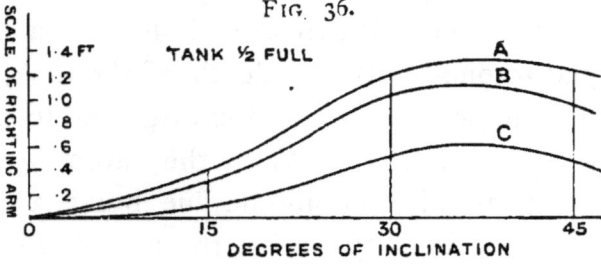

FIG. 37.

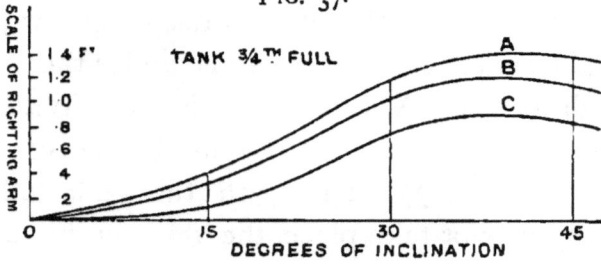

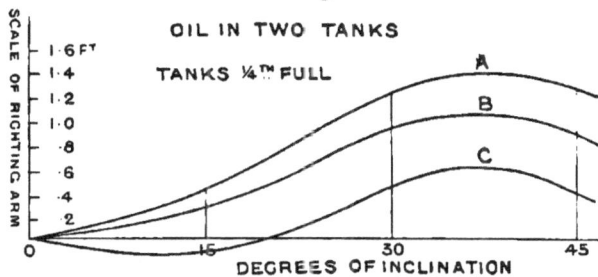

FIG. 38.

OIL IN TWO TANKS

TANKS ¼TH FULL

SCALE OF RIGHTING ARM

1·6 FT
1·4
1·2
1·0
·8
·6
·4
·2

A
B
C

DEGREES OF INCLINATION

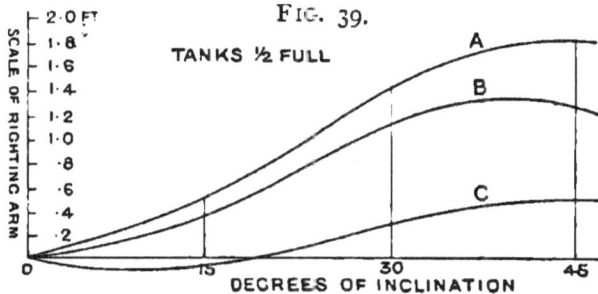

FIG. 39.

TANKS ½ FULL

SCALE OF RIGHTING ARM

2·0 FT
1·8
1·6
1·4
1·2
1·0
·8
·6
·4
·2

A
B

C

DEGREES OF INCLINATION

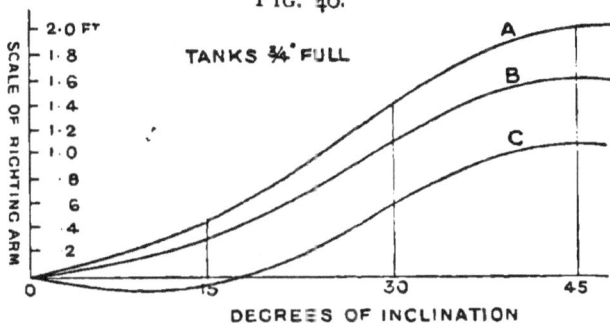

FIG. 40.

TANKS ¾ FULL

SCALE OF RIGHTING ARM

2·0 FT
1·8
1·6
1·4
1·2
1·0
·8
·6
·4
·2

A
B
C

DEGREES OF INCLINATION

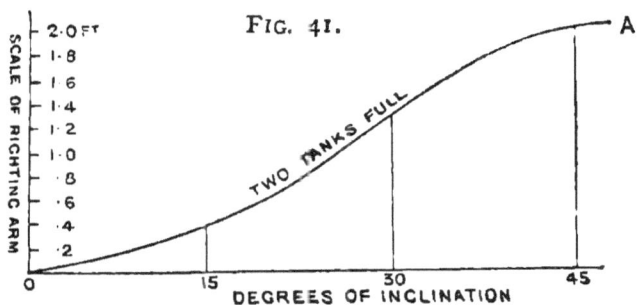

FIG. 41.

SCALE OF RIGHTING ARM

2·0 FT
1·8
1·6
1·4
1·2
1·0
·8
·6
·4
·2

TWO TANKS FULL

A

DEGREES OF INCLINATION

Figs. 38 to 41 give the curves of righting arm when two tanks are filled simultaneously, instead of one, as in the preceding example. A, Fig. 38, is the curve of righting arm when the two compartments are one-fourth full, the cargo being supposed not to shift ; B is the curve of righting arm on the assumption that the cargo shifts, and that the middle-line bulkhead is oil-tight; and C is the corresponding curve on the assumption that the middle-line bulkhead is perforated. In the latter case it will be seen that the vessel does not stand upright, but inclines away through $19\frac{1}{4}°$ before attaining a position of equilibrium. With the compartments half full the corresponding curves are A, B, and C, Fig. 39. Here, in the most critical condition, the inclination at which the vessel comes to rest is 18°. With the compartments three-fourths full the curves are A, B, and C, Fig. 40, and with the compartments quite full they unite, forming A, Fig. 41. It will be noticed in the above examples that when the two tanks are partially filled, the middle-line bulkhead being perforated, while the vessel inclines away from the upright through a considerable angle before coming to rest, the reserve of stability remaining is but small, and if it were attempted to fill three tanks simultaneously under the above circumstances the vessel would in all probability capsize.

The curve on Fig. 42 shows the manner in which the inclination varies as the depth of the oil in the two compartments changes under the above circum-

stances ; and it will be of interest to trace the vessel's
behaviour during the filling of the tanks. Assuming
her to be upright when the oil begins to pour into the
tanks, she will continue upright until the oil attains a
definite height.   Owing to the low position which the
incoming oil occupies, the common centre of gravity of

FIG. 42.

DIAGRAM SHOWING INCLINATION
OF VESSEL DURING FILLING OF TWO
COMPARTMENTS

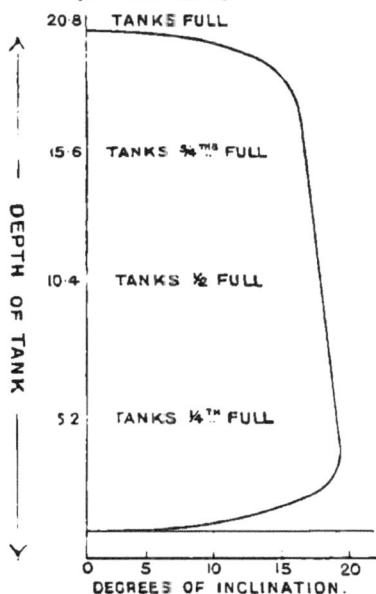

hull and cargo falls, but m rises rapidly as the oil in
the tanks rises, because of the rapidity with which the
free surface grows, and M usually falls.   When m and
M coincide the vessel is in neutral equilibrium.   This
occurs when the oil has risen to a height of 15 inches
in the case under consideration.   Immediately after-

wards, as the oil continues to pour into the tanks, m rises above M, and the vessel inclines rapidly, reaching a maximum inclination of $19\frac{1}{2}°$ From that point onward the vessel begins to right herself, but very slowly, until the oil surface rises to near the level of the deck edge.   Thereafter the inclining effect of the transfer of the wedge of oil from one side to the other diminishes rapidly, and the vessel moves fast towards the upright until at last m passes below M again, when the oil is only a few inches below the crown of the tank, and she becomes upright.

76. In respect of the demands made upon her stability, the oil-carrying steamer differs greatly from the ordinary cargo vessel.   With ordinary cargoes a vessel's stability is most tasked when she is rolling amongst waves at sea.   She there requires a sufficient range to provide a margin against the heave of the sea, the inclining effect of the wind pressure, and against that prolific cause of disaster, the shifting of cargo. Whilst loading or discharging in dock, or alongside a wharf, no great amount of initial stability is necessary, and if a list does take place through an unequal distribution of the cargo, it becomes at once so apparent that the necessary steps may be taken by regulating the stowage to restore her to the upright. An oil steamer is, however, in a somewhat different position at sea.   So long as the tanks remain full, shifting of the cargo is made impossible, but on the other hand, if from any cause a subsidence in the level of the oil takes place, shifting does occur with a

celerity unparalleled with non-liquid cargoes. Wheat, for instance, does not shift when at rest until its surface has a slope of some 26 deg., and even when at sea, although shifting takes place at a somewhat smaller inclination, owing to causes whose effects are discussed and measured in the paper on the shifting of cargoes to which reference has been made,* the inclination still remains considerable. With free oil or other liquid, however, the slightest inclination of the vessel causes a corresponding change of surface. The most obvious way of minimising the danger from this source, next to careful workmanship and sufficiently close riveting, is to divide the cargo hold so as to restrict the area affected in the event of leakage taking place. The latter end is most effectively attained by fitting a practically oil-tight middle line bulkhead, which, as has already been shown, has a most important effect in conserving a vessel's stability. In that case, however, it is desirable to provide appliances for filling or discharging tanks lying opposite each other at the same time. If this be not done serious inclination may result owing to the shifting of the common centre of gravity of hull and cargo away from the middle line as already explained. Numerous transverse bulkheads too, while they have no such effect as the middle-line bulkhead in respect of transverse stability, limit the area over which leakage can extend, and reduce the

* See Paper "On the Shifting of Cargoes," by Prof. Jenkins, read at the Inst. Naval Architects, Session 1887.

straining effect which longitudinal motion of the cargo would cause.

77. But while the oil steamer at sea is particularly safe as regards stability, so long as the level of the oil does not fall, and can indeed afford to prosecute her voyage with a smaller curve than most other classes of vessels, it is during the process of loading or unloading that usually the need for large initial stability and effective longitudinal subdivision is most felt. When the vessel is brought alongside the wharf and communication effected between the tanks and the reservoir on shore, the greater the number of tanks that can be filled simultaneously the sooner is the loading completed, and the earlier is she able to put to sea. Similarly as to the discharge of the cargo. As we have seen, the condition which limits the number of tanks that may be so treated is one of stability only, as alike during loading and discharging the tanks soon become partially filled, and if too great a number are being so treated the vessel will cease to remain upright, and may incline over to a considerable angle before attaining a position of equilibrium. It might be thought that one good way of increasing the number of tanks that may be filled simultaneously is to make the vessel very broad. Such a result would no doubt be attained in that way, since M rises in the vessel with increase of breadth more rapidly than m ; but oil steamers have as a rule quite enough initial stability when laden, and any important increase of breadth would lead to great stiffness, and therefore

heavy rolling, which is very undesirable with this class of vessel.

78. In practice the number of tanks that may be simultaneously filled increases as the loading proceeds, and the number that may be simultaneously discharged decreases as the vessel rises out of the water. It is not difficult in any given case to draw up regulations insuring that a vessel may be loaded or discharged with the greatest rapidity consistent with safety.   Indeed, in this respect, as well as in respect of her stability at sea, the problem of the oil steamer is much simpler than that of the ordinary cargo-carrying vessel.   The latter is engaged as the market demands, seldom carrying two cargoes alike, and frequently shipping mixed cargoes, the scattered arrangement of which it is impossible for the naval architect to anticipate by calculation in any estimate of stability.   Rough rules to the effect that a certain proportion of the whole cargo should be stowed in the hold and the remainder in the tween-decks, as for instance two-thirds in the hold and one-third in the tween-decks, although common with stevedores and others, are very unreliable unless due allowance be made for the nature of the cargo and the method of stowage.   Moreover, under such a rule, no cognisance is taken of differences in the breadth of vessels having the same depth, and in that way broad vessels may be made too stiff, and narrow vessels too tender.   The oil steamer is in the position of always carrying a definite amount of homogeneous cargo-oil on the

outward voyage, and water ballast on the inward, and is therefore capable of having her stability determined with great nicety in each of the conditions in which she is liable to be placed.

In concluding his remarks, Prof. Jenkins advises that the following information should be given to the captain by the designer of the vessel.

1. The number and positions of the tanks to be simultaneously filled at the loading port, from the commencement of the loading to its completion.

2. The number and positions of the tanks to be simultaneously emptied at the end of the voyage, from the commencement of the discharge to its completion.

3. The number and positions of the tanks to be filled with fresh water for the return journey, so as to give a proper trim and sufficient stability, together with the order in which they should be filled.

4. The order in which the fresh-water tanks should be pumped out on arrival at the loading port.

79. Carrying oil or any liquid at sea in bulk is attended by different conditions to those which govern ordinary cargoes. Consequently, the handling of petroleum vessels must be governed by different rules.

It hardly would appear necessary to point out that certain precautions are necessary in the case of oil which do not obtain in the case of ordinary cargoes. When a ship is moored in dock we like to see her loading or discharging from all hatches at once, and what with double derricks and winches it is really astonishing how cargoes of 3000 and 4000 tons are handled with such speed and safety ; but the captain who seeks to fill or empty all his oil tanks at once will not only be doing a very stupid thing but will also probably incur considerable risk and danger to the ship. We once saw a man attempt this ; he did not do it again.

80. In petroleum vessels much diversity of design exists ; some are the result of careful and scientific design, while some few others are, as petroleum vessels, fearful and wonderful examples of "how not to do it." In the earlier vessels many contrivances were adopted, such as a system of cylinders and elaborate piping arrangements. The object being to divide the cargo into small lots, so as to prevent leakage and promote safety.

81. As we have insisted, too much importance cannot be attached to keeping the tanks at sea quite full. The whole safety of a petroleum vessel at sea consists in having the cargo so stowed that the C.G. does not shift.

82. On this subject Mr. Martell says, in his able paper :—

" Let it be supposed that the vessel in Fig. 43, which contains oil with a free surface O O_1, becomes inclined through a small angle O S H. The wedge of oil O S H is transferred to the position O_1 S H_1, and consequently the common centre of gravity of hull and cargo moves away from its former position G in the middle line to a new position G_1. The centre of buoyancy also moves from B to B_1 in the ordinary way. We shall have, therefore, the resultant upward pressure of the water, equal to the weight of the laden vessel, acting vertically through B_1, and an equal weight downwards through G. The distance A M between the points in which the lines of action of the two forces cut the middle line of the vessel measures the righting moment at small angles of

FIG. 43.

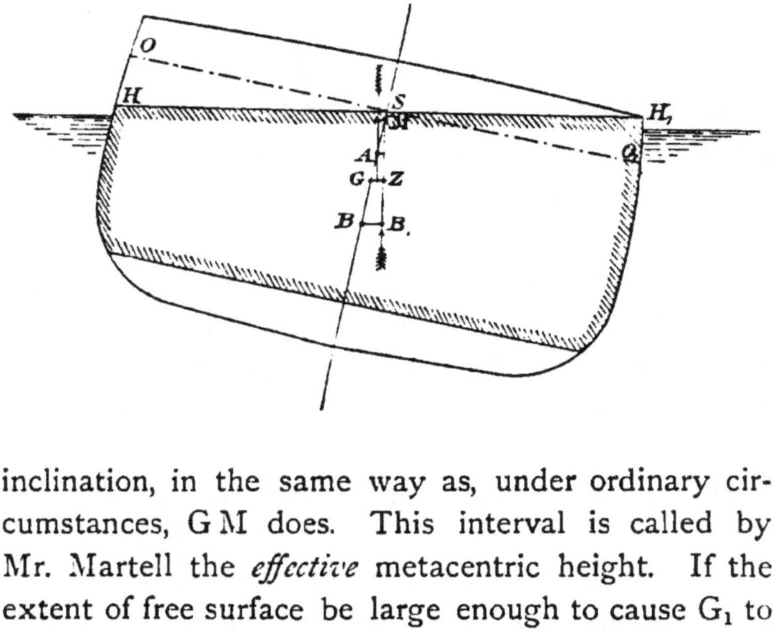

inclination, in the same way as, under ordinary circumstances, G M does. This interval is called by Mr. Martell the *effective* metacentric height. If the extent of free surface be large enough to cause G_1 to pass beyond the vertical through B_1, the vessel will be unstable in the upright condition. If the vessel has a ballast tank beneath the oil cisterns in which leakage to any considerable extent can collect, an additional wedge of oil is transferred from one side to the other

K

as the vessel inclines, as shown in Fig. 44, resulting in a further decrease in the *effective* metacentric height.

To illustrate the importance of this point he takes the case of a "Three-deck" vessel 250 feet long, by 33 feet broad, with a load displacement of 3500 tons, to have a midship compartment 30 feet long extending from side to side. If the level of the oil in such

FIG. 44.

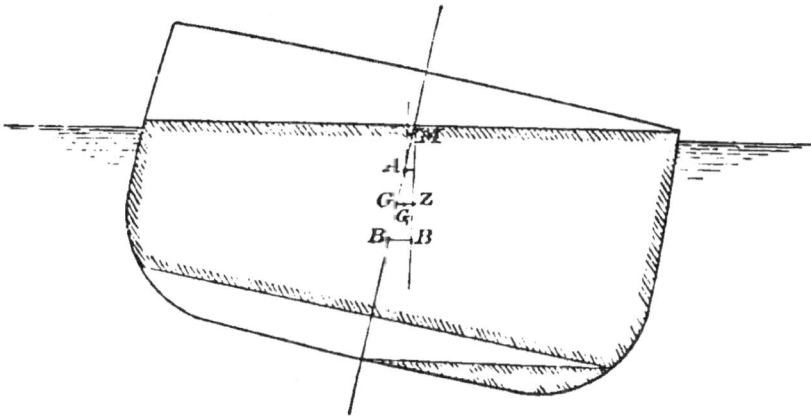

a compartment were slightly lowered, the *effective* metacentric height would be reduced by as much as 7 inches, and if the leakage were to collect in the bottom of the vessel over the same area, we should have, in the case of a flat-bottomed vessel, nearly double this decrease in metacentric height. It will be easily understood that such a loss of metacentric height might in practice be attended by serious consequences, and hence it is very desirable to minimise the area over which leakage at any one point would have effect, by subdividing the hold space, as well as by making good any loss due to leakage. By fitting a longitudinal middle line bulkhead, the loss of *effective* metacentric height by a fall in the level of the oil over the same area is reduced to one-fourth ; and by subdividing the oil-carrying space by numerous transverse bulkheads, the extent of the free surface in connection with leakage in any one compartment becomes reduced, and the consequent danger due to a loss of initial stability correspondingly diminished.

83. Closely connected with this part of the subject is the question of the proper proportion between the breadth and depth of a vessel designed to carry petroleum in bulk. If the breadth be great in relation to the depth, and the vessel has, in consequence, great initial stability, and is, therefore, a heavy roller, a depreciation in the level of the oil will be followed by an agitation tending to cause the compartment to strain and leak. If, on the other hand, the vessel be so narrow as to have only a small metacentric height, any fall in the level of the oil in one or more compartments will be liable to render her unstable in the upright condition.

The suitable relation between the breadth and depth can, of course, only be settled in each case in connection with the proposed arrangement of tanks, but it is obvious that vessels in which the oil extends to the inside of the plating of the bottom, and in which the crown of the tanks is fitted some distance

K 2

below the upper deck, as well as those which carry only a limited amount of oil in the tween-decks, should be made relatively narrower than vessels which have the tanks so arranged that the position of the common centre of gravity approximates to that usual with ordinary cargoes.

As regards the proportions that should exist between the breadth and depth of a vessel designed for carrying oil in bulk :—If the beam is great compared with the depth, the initial stability will be large, and the vessel will roll a good deal in a sea-way ; and if there is any leakage, the oil running from side to side in the tanks will be agitated, and give off dense volumes of vapour. Also, by a tendency to excessive rolling, great stresses are brought upon the ship, which may eventually open the seams and start the caulking of the tanks. On the other hand, a narrow vessel, while steadier in a sea-way, has as a rule little initial stability but a larger range, provided always that the tanks are quite full. Any leakage, as has been explained, is followed by a serious loss of metacentric height. From what has been said, it will be gathered that the writer is opposed to ballast tanks for oil-steamers —that is, when the crown of the tank forms the bottom of the oil tank.

84. Considerable difference exists upon the question whether ballast tanks should or should not be fitted for petroleum-carrying vessels, and, if fitted, the method to be adopted. At present we have some vessels with ballast tanks of the ordinary construction,

in others no ballast tanks at all are fitted.   Again, in
the more recent ones, the ballast tanks are situated in
the ends, and the writer is inclined to think this the
best method.   It will be observed that in the vessels
designed by Sir E. J. Reed and Messrs. Flannery and
Blakeston, this method is adopted, and the writer
prefers to err in such good company rather than advocate any other principle.   Some objection may be
taken to fitting ballast tanks in the ends ; but when it
is recollected that the best ship of any class is simply
but a compromise in which several good qualities have
to be mutually sacrificed, it will be seen how impossible it is, in the case of petroleum vessels, to design a
vessel that will at once satisfy all conditions.   If we
examine the curve of loads in tons per foot of length
for a vessel fitted with ballast tanks in her ends, it will
be seen that when light and when laden very considerable difference of stress in the structure takes
place.   In fact it is simply for reasons of expediency
and safety that the principle is adopted.

85. Whatever system be adopted, it is necessary
that the workmanship should be of the highest class,
and careful supervision during building is necessary.
It must not be forgotten that when in a sea-way the
weight of the cargo in a ship is increased as much as
20 per cent. when in the trough of a wave, and
diminished by a corresponding amount when on the
crest, thus bringing severe stresses on the tanks.   In
the writer's opinion oil tanks should be a " boilermaker's job," and as such it should be called in the

specification.   The riveting should receive special
attention.   The holes should be perfectly fair with
each other.   In the construction of the tanks the

FIG. 45.

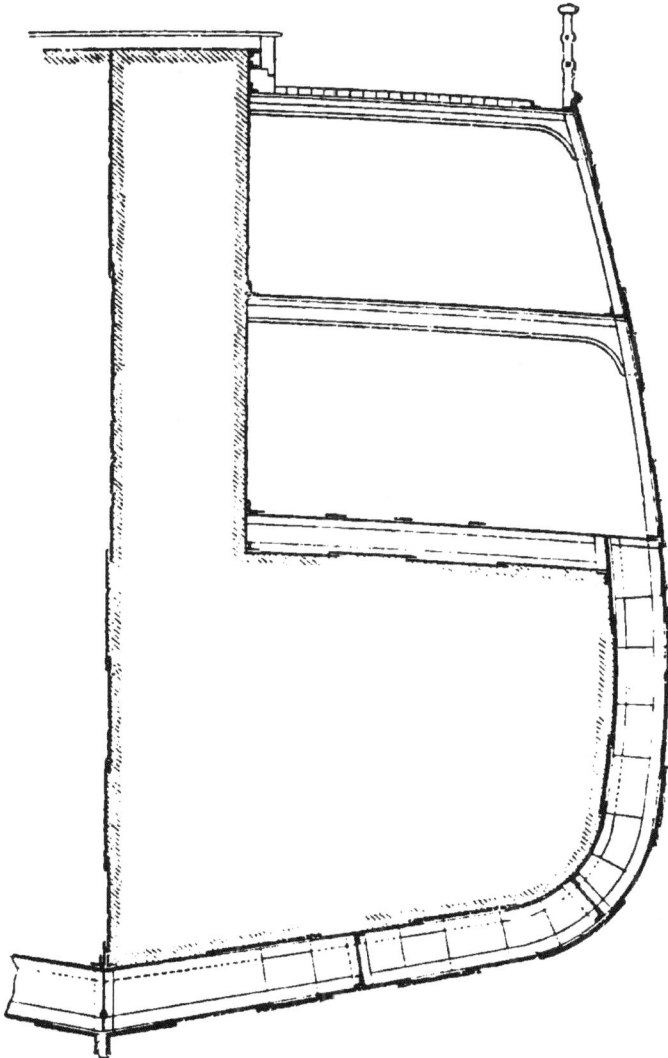

rivets should be not less than $\frac{3}{8}$ inch diameter, and spaced not more than $2\frac{1}{4}$ inches apart between centres. In the shell plating the rivets should be $\frac{7}{8}$ inch, and spaced not more than $2\frac{3}{4}$ inches between centres.

86. We will now give a short description of the various types of oil-carrying steamers.

No useful purpose will be served by describing the various methods for carrying oil in bulk as adopted by sailing vessels, as it is not likely that this branch of the business will ever be of very much importance, and those sailing ships that may hereafter carry oil in bulk will be fitted much in the same way as steamers.

87. Figs. 45 and 46 are a transverse and longitudinal section of the *Vaderland*, *Nederland*, and *Switzerland*. This design has much to recommend it. The vessel has a good tumble-home that sailors will appreciate. The arrangement of the tanks is good, and, being built on the

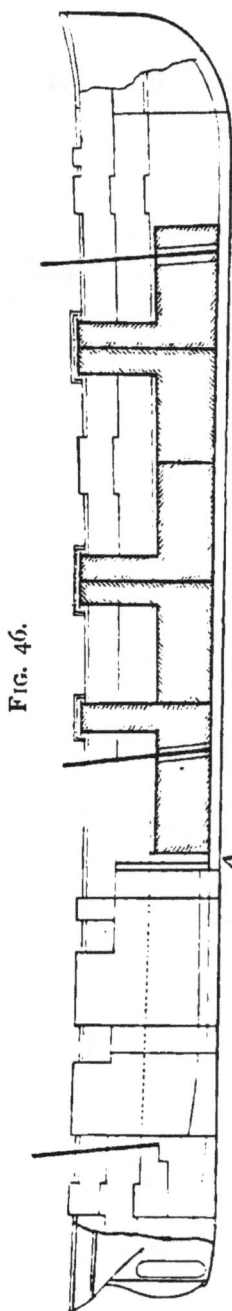

Fig. 46.

cellular system, the advantages of water ballast are
obtained with very much less risk of danger from
leakage of oil.

FIG. 47.

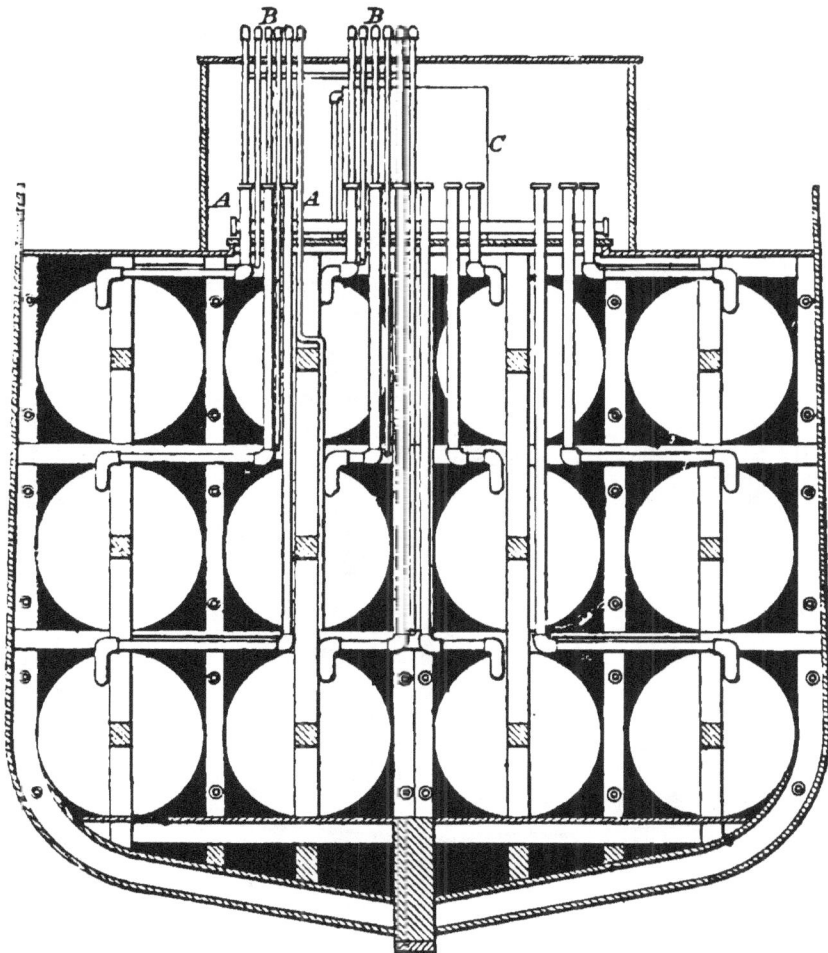

88. Fig. 47 is a transverse section of the sailing
vessel *Crusader*. We give this example as it is
being adopted for steamers in the States.

89. Fig. 48 is a transverse section of the *Fergusons* steamship.

This arrangement of tanks is not likely to be copied.

FIG. 48.

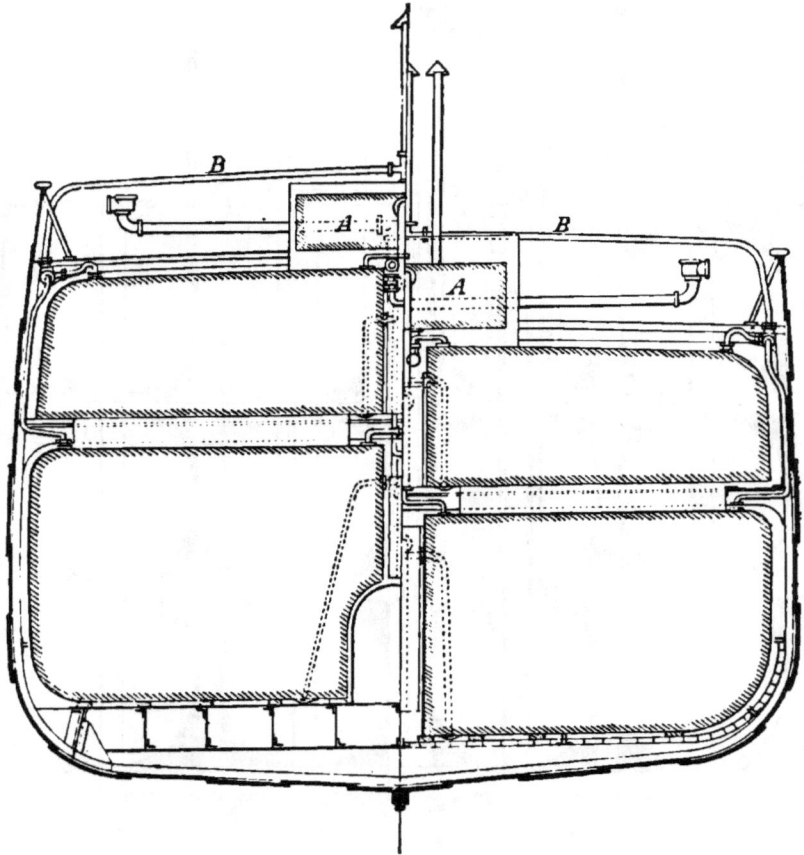

90. Fig. 49 is Mr. Swan's earlier arrangement, as carried out in the *Gluckauf* and other vessels. The hold is divided by a middle line longitudinal bulkhead, and also by a series of transverse bulkheads,

into compartments, and to each compartment is fitted
one or more trunkways or expansion and filling
tanks.

It will be observed that the ballast tank is conical
in section, and by this arrangement the oil tanks can
be drained very effectively. In the event of there
being any leakage into the ballast tank, the latter is

FIG. 49.

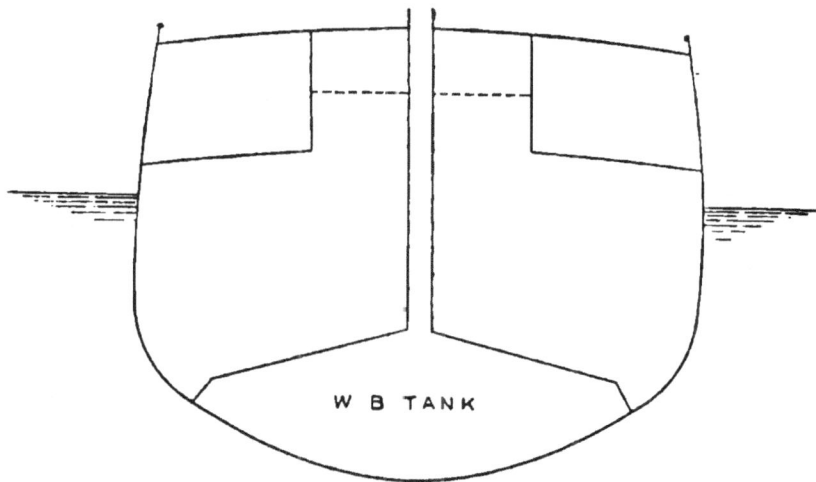

SECTION THROUGH TRUNKING.

filled with water, and the oil floats up the midship
trunkway, from whence it can be pumped back into
the main tank. This would hardly be attended with
good results if the vessel was carrying refined kerosine,
as cloudiness would result, which might cause the lot
to be thrown on the owner's hands, or would entail a
considerable reduction in the freight paid. We believe
that Mr. Swan does not, however, advise the use of

ballast tanks, and in his recent vessels, such as the *Lumen* and *Lux*, there are no ballast tanks excepting on the ends.

Another plan for improvements in the construction of "navigable vessels for carrying liquids in bulk including cargoes of a volatile character, such as petroleum, turpentine, and the like" has been proposed by Mr. Swan, of the firm of Sir W. G. Armstrong, Mitchell & Co., Limited. By Mr. Swan's method the hold is divided by a middle line longitudinal bulkhead, and further sub-divided by a series of transverse bulkheads into compartments bounded at the top either by the vessel's deck or a specially fitted platform. As in some of the Caspian steamers, the oil extends to the skin of the vessel, and to each compartment are fitted one or more trunkways, either circular in section or of any other convenient form, partially filled to ensure that the corresponding compartment is full, and also to provide for the contraction or expansion of the oil. A number of ways in which the trunkways may be fitted are suggested by Mr. Swan. One feature of the method is the arrangement for recording the height at which the liquid stands in the trunkway. For this purpose a piston is fitted to the trunkway and floats on the liquid, a graduated rod attached to it passing up through a stuffing box in the cover. A small pipe in the cover of the trunkway admits of the escape of vapour. As in other plans, by a suitable arrangement of pipes each compartment is capable of being

separately filled or emptied. To provide for any
leakage through the end bulkheads of the com-

FIG. 50.

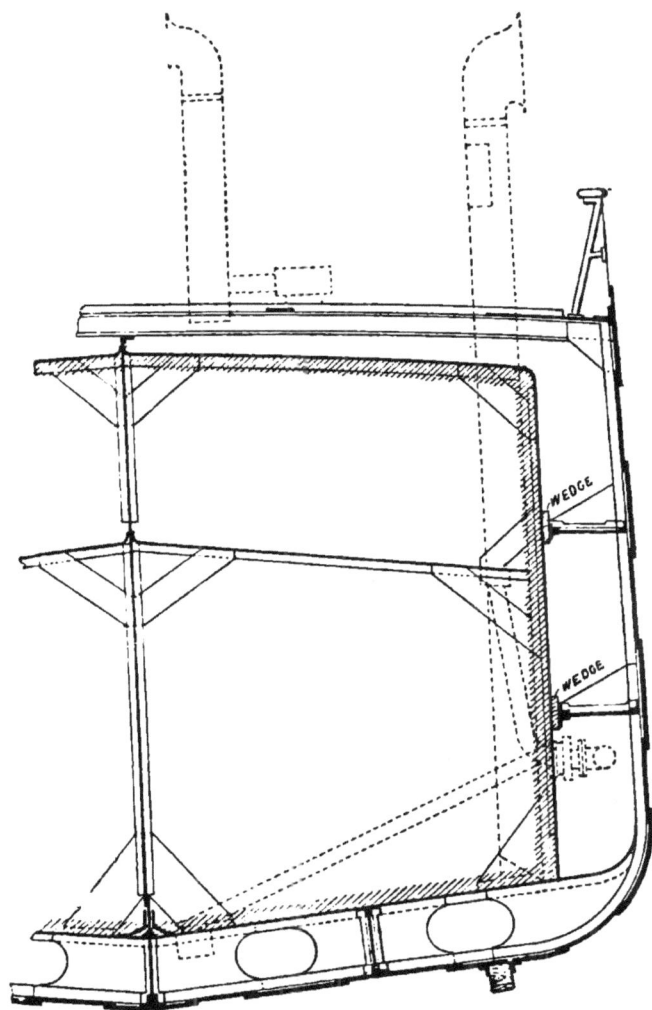

partments a well is formed on the fore side of the
boiler space, and arrangements are provided to

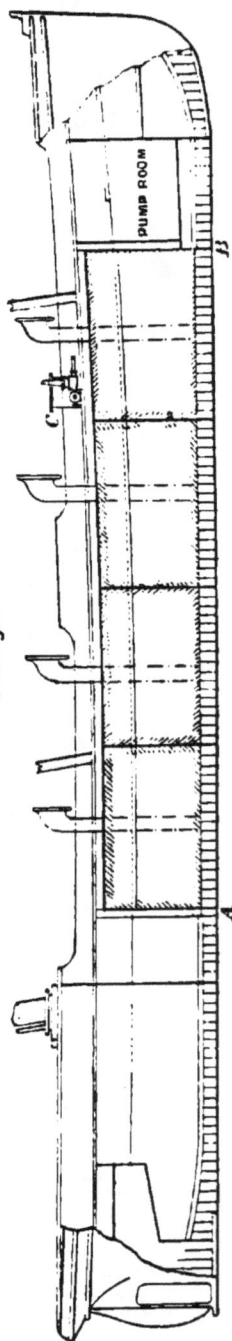

enable the liquid to be pumped back into the trunkways.

91. Figs. 50 and 51 illustrate the *Sviet*, built for the Russian Steam Navigation Company by the Motala Company, who have had great experience in the building of vessels intended to carry petroleum in bulk. She has been specially designed to convey the oil from Batoum to European ports, and differs in her construction from those already described. Her engines are placed aft, and double bulkheads, marked A on Fig. 51, forming a "water bulkhead," are fitted on the fore side of the cross bunker. Forward of this, for a length of 140 feet, the vessel is divided by a middle line longitudinal bulkhead, and by several transverse bulkheads extending to the upper deck. In each of the compartments so formed a cistern is constructed, the sides of which are built within the sides of the vessel, leaving a passage of about 3 feet. Similarly, a space of about

FIG. 51.

20 inches is left between the upper deck and the
top of the cistern. The plating of the top of
the cellular bottom with which the vessel is fitted
forms the bottom of the cisterns, which are further
divided by a steel watertight flat into an upper and
a lower tier. There are 15 such cisterns in all.

92. Figs. 52 and 53 are transverse and longitudinal
sections of the *Bakuin*, owned by Mr. A. Stuart, of
London.

FIG. 52.

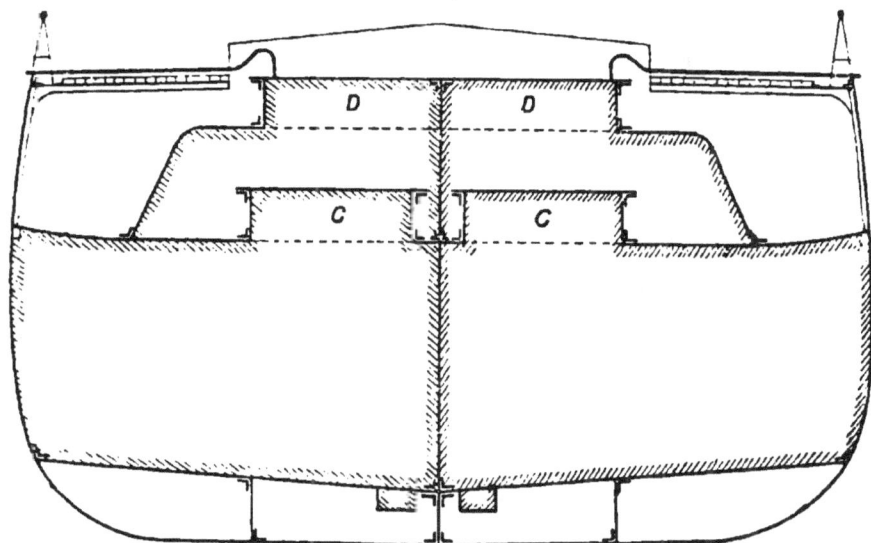

As shown on the sketch of midships section, the
Bakuin has a cellular bottom, the crown of which
forms the floor of the oil tanks. Above the cellular
bottom, to the height of the tween-decks, the oil
extends to the side, as in some other plans. In the
tween-decks, and of the shape shown on the section,
are built a number of additional oil compartments.

They do not extend either to the side of the vessel
or to the deck above, and it is claimed for this plan
that while the oil in the main hold can never reach
a high temperature, owing to the immersion of the
vessel, the tanks in the tween-decks, by being so
formed, are kept at a much lower temperature in hot
climates than if they extended to the sides. In the
event, too, of injury to the hull at this part by collision
or otherwise, the tanks would, under ordinary circumstances, escape injury.

As in other modern vessels, the machinery is placed
aft, and a double bulkhead, marked A on Fig. 53, is
fitted before the boiler space, and another at the fore end
of the foremost oil compartment. The hold is further
separated into two distinct divisions by an additional
pair of adjacent transverse bulkheads (B), as shown on
the longitudinal plan. The object of this arrangement
is to allow of oils of different qualities being carried
on the same voyage without any danger of their
mixing. Arrangements have also been made by
which, with oil in the hold, other descriptions of
cargo may be carried in the tween-decks. With this
object, expansion tanks (C) formed on the middle
deck communicate with the cisterns in the hold, and
are capable of being closed, air pipes being fitted to
pass through the cover and above to the upper deck.
Additional expansion tanks (D), built on the roof of
the tween-decks cisterns, within the area of the
upper deck hatchways, are for use when both the
tanks in the hold and tween-decks are filled.

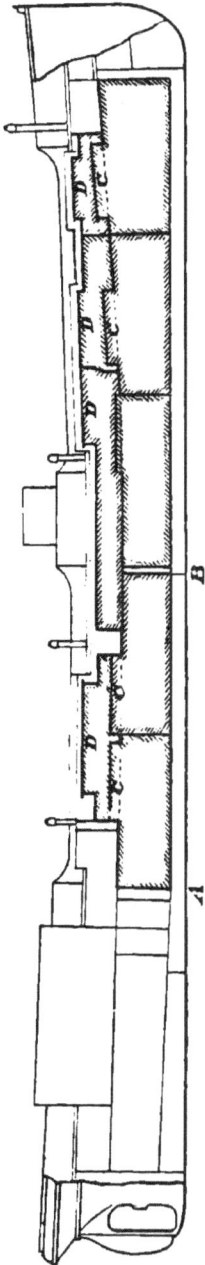

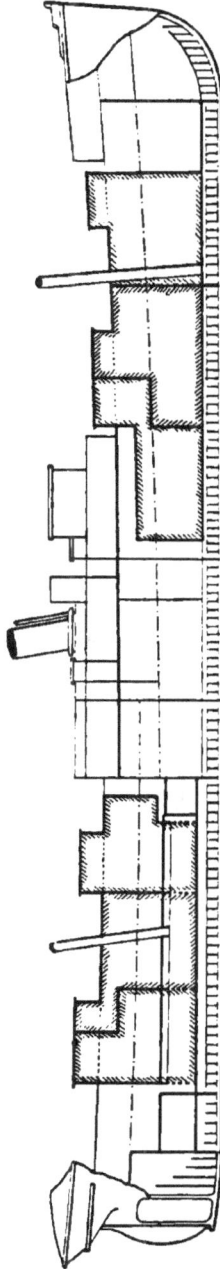

Fig. 53.

Fig. 54.

The whole of the valves which regulate the filling or emptying of the tanks are conveniently placed in the engine house, and by means of an arrangement of floats, connected with wires, the level of the oil in each tank can be ascertained at the same place. Great care appears to have been taken in the construction of the *Bakuin* to avoid all possible sources of risk from fire. She is lighted by electricity; the cabins are heated by steam instead of by fires; and the cooking is or was done by steam. It is estimated that this vessel will carry 1950 tons of oil; and the pumping arrangements are such that, with proper facilities for discharge, the cargo can be emptied in about twelve hours.

In the writer's opinion the arrangements here described are not the best, and, bearing in mind our remarks on ballast tanks, we do not think that this vessel will be copied extensively; on the other hand the proportion $\dfrac{D}{B}$ is good. This vessel carries her cargoes very well.

93. Figs. 54 and 55 are a longitudinal and half-transverse section of the *Chigwell*—the vessel has been "converted." The peculiar arrangement of the tanks will be noticed, but in this case, as in other "converted" vessels, the builders had to do the best they could. The arrangement of the tanks is not such as would be adopted on a new vessel. It is only fair to say that this vessel is exceedingly strong, and carries her cargoes very well indeed. Her owner, Mr. A. Stuart,

L

of London, owns several "converted" tank steamers, as well as others of the most recent design.

94. It is hardly necessary to emphasise the employment of iron or steel in decks and deck erections of

FIG. 55.

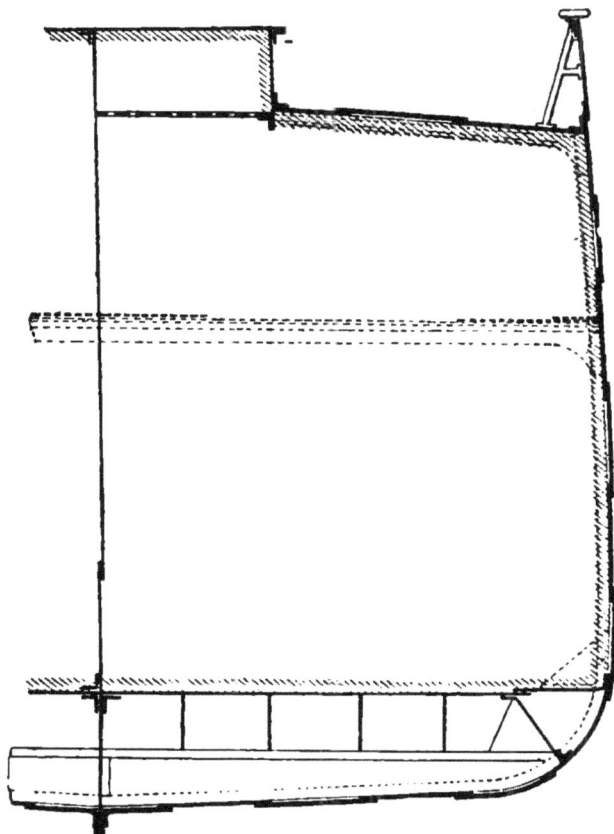

petroleum-steamers. Soft woodwork should be religiously avoided, as it soon becomes impregnated with petroleum vapour. It sometimes is necessary to pump the refuse of a tank on deck, and if the decks are of wood, the oil is absorbed, and an accidental flame would quickly produce disastrous results. In the latest types of petroleum steamers no woodwork is used at all, more than is absolutely necessary.

95. Fig. 56 is a longitudinal section of the *Charlois*, a very fine vessel, designed by Messrs. Flannery and Blakiston, of London and Liverpool. The writer has been favoured with a description of this vessel, which may be of interest, inasmuch as she embodies the latest practice, and in the writer's opinion the design is worthy of being imitated.

This vessel is 310 feet long, 39 feet beam, and 25 feet 3 inches deep, and is capable of carrying upwards of 3500 tons of petroleum, besides bunker coal, on a moderate draft.

The principal feature in this vessel, as in all the more recent steamers built for this trade, is the fact that the oil is carried to the skin of the ship, that is to say, that the ship's structure itself forms the necessary receptacle for

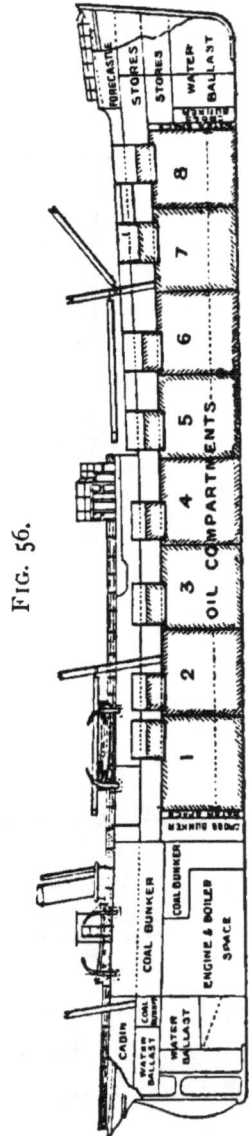

FIG. 56.

carrying the oil, being subdivided by bulkheads into compartments of moderate size. In the majority of the earlier vessels built for carrying oil, separate tanks were built into the ship for carrying the oil ; and this was not only much more expensive, but very much more dangerous, owing to the number of empty spaces left for the accumulation of gas. A similar objection holds good against the introduction of a double bottom as adopted in some more recent vessels, as in practice it is found almost impossible to avoid leakage from the tanks to the double bottom, with its attendant risks ; and as the advantages of a double bottom are not so pronounced with this class of vessel as in ordinary cargo boats, there is not sufficient inducement to incur the risk of adopting it. The designers of the *Charlois* (Messrs. Flannery and Blakiston) clearly recognised the disadvantage of this, and in this vessel the oil is carried not only to the sides but the bottom of the ship, her construction being therefore of the simplest possible. The requisite subdivision into eight tanks of moderate size is obtained by the introduction of nine thwartship bulkheads, which are very heavily stiffened and made extra thick to withstand the pressure due to any one tank being full while the others are empty. In addition to these there is a longitudinal bulkhead running the entire length of the oil compartments in the centre of the ship, which further subdivides each tank into two. Wells, or water spaces, are formed at each end of the oil compartments, which are filled with water when the vessel is loaded with oil, and thereby isolate the oil from the rest of the ship and boiler room, to prevent risk of fire. As will be seen, each tank is provided with a smaller tank above, running up through the tween-decks to the upper deck, which is fitted to allow for the expansion and contraction of the oil, due to difference in temperature, without permitting the oil to ever fall below the level of the top of the tank proper, which is essential to the vessel's stability at sea. It is usual to carry these expansion tanks about half full of oil. The expansion tanks also serve the purpose of giving access to the tanks, proper manholes and Jacob's ladders being provided. A special feature in this vessel, which the designers had particularly in view, is the fact that she could, with very slight alterations, be used for ordinary cargoes ; the expansion tanks being arranged conveniently, and of extra size, for this purpose ; this is, we think, an important point. The machinery and boilers are placed close aft, and clear of the oil compartments, and the saloon and officers' and engineers' cabins and gallery, are abaft this, and therefore well clear of the tanks. The crew are berthed in the forecastle, and there is a long bridge amidships, with a shade or awning deck, connecting it to the poop. The shade deck was specially introduced by the designers to make the vessel more seaworthy, as she is employed in the Atlantic trade, and this considerably reduces the amount of exposed deck. The internal fittings of this vessel are most complete, and, as a

further precaution against fire, she is lighted throughout by the electric light on the incandescent principle,
the engines and dynamos being placed in the engine
room directly under the control of the engineers.
There are also steam heaters for all the cabins and the
crew.   Two powerful pumps are fitted in the tween-
decks with very complete piping arrangements, each
pump being capaple of discharging the entire cargo of
oil in thirty hours.   As pointed out, there is no double
bottom for water ballast, but tanks are provided at
both ends for trimming purposes.   For ballasting the
ship when light, two or more of the oil tanks are run
up with water, special means being provided for this
purpose, and when so laden the vessel is much steadier
at sea than if carrying ballast in the ordinary double
bottom ; and, as these vessels have to make one out of
every two trips across the Atlantic, light ship, this is
very important.

96. Fig. 57 is a longitudinal and transverse section
and deck plan of the steamers *Era* and *Oka*.   These
steamships were designed by that eminent naval architect Sir E. J. Reed, and are without doubt the most
perfect specimens of design for their purpose.   The
question of stability in these vessels has received that
special attention which it is well known Sir E. J. Reed
bestows on all his vessels, and in these two it is
almost an impossibility for any trouble to arise.   The
Era is 271 feet long, 37·2 feet beam, and 22·5 feet
deep.   She is 1195 tons registered and 1851 gross,
and carries about 2000 tons of oil.   Her water ballast

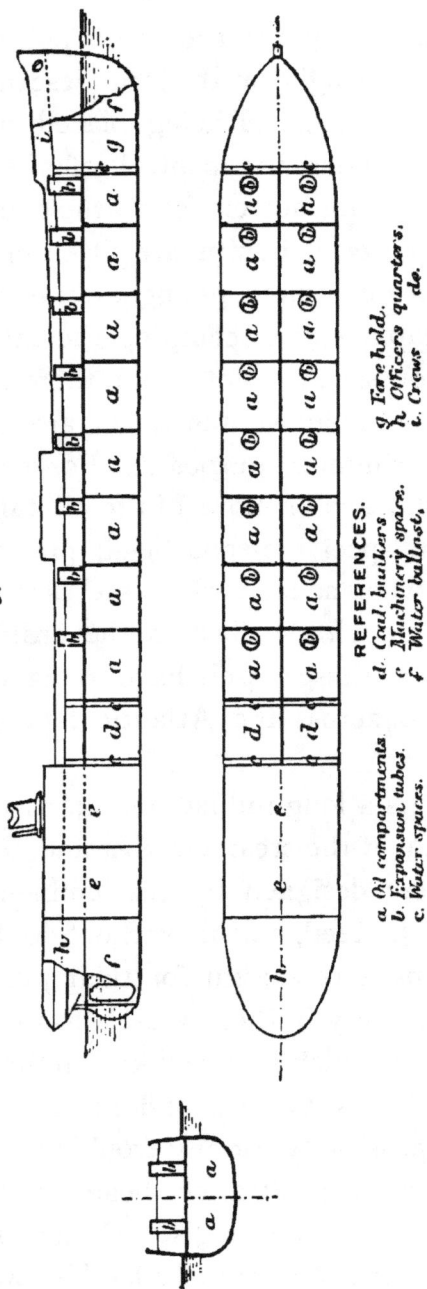

FIG. 57.

REFERENCES.

a. Oil compartments.
b. Expansion tubes.
c. Water spaces.
d. Coal bunkers.
e. Machinery space.
f. Water ballast.
g. Fore hold.
h. Officers quarters.
i. Crews do.

is carried in her ends. The forepeak tank contains
46 tons, and the after peak 55 tons. Additional
immersion can be obtained when she is light by filling
one or more of the oil tanks. A very good feature in
this and the following vessel is the arrangement for
carrying the cargo ; it will be observed that with the
exception of that contained in the expansion tubes,
all the cargo is below the load water line, thus the
stability of these two vessels is perfectly assured.
The large tween-decks can carry a very large amount
of coal, and thus the *Era* offers an advantage to her
captain which very few vessels do : and that is, it is
in his power to so counterbalance the weights as to
obtain the proper amount of G.M. The " stability
due to form " in these vessels is very large, and we
doubt whether they could be upset. This cannot be
said for all tank steamers.

The *Oka* is 334·4 feet long, 44 feet beam, and
24·9 feet deep, 1995 tons register, and carries about
3000 tons of oil in the very small draught of 16·9 feet.
In Fig. 57, the tubes marked *b* are expansion tubes,
and their total capacity is equal to 2 per cent. of the
volume of the cargo. They are fitted with gauges and
pipes, by which a perfect level of oil can always be
maintained. The remainder of the explanatory notes
on the sketch speak for themselves. The method by
which the vessels are filled and emptied of their cargo
is not shown, but ample pumping and piping arrangements are fitted to each vessel, by which this is done in
the most efficient and rapid manner. The spaces marked

"*f*," which are appropriated for water ballast, are so designed as to keep the vessel on a level keel when she arrives in England fully loaded with oil, but with her bunker coal consumed. The vessels are of course lighted throughout by electricity. Each separate tank is provided with ample means of ventilation. Steam is used for heating and cooking.

In the writer's opinion the *Era*, *Oka*, and *Charlois* fully solve the problem of carrying oil in bulk, and, providing due intelligence is used (it is not always) by the captains and officers of these boats, no danger can possibly arise.

97. We should very much have liked to have given the curves of displacement, stability, and metacentres, for these last three vessels, but somehow there is an objection on the part of builders and designers to give this information. This is to be regretted, as it assumes that there are points which had best not be known, and the writer contends that in all cases there should be deposited with some public authority full particulars of the qualities of each vessel, so that in case of loss or damage there shall be no difficulty in arriving at the truth. It is a matter of deep regret—and one which will, the writer trusts, be remedied—that in inquiry cases as to the loss of a vessel the technical information is seldom forthcoming, and the court is left to make guesses at truth, more or less (generally less) correct. At the same time it must be confessed that the composition of the average court is not such as lends itself to the ready interpretation of stability

curves. The look of blank astonishment which the
writer once saw on the face of a nautical assessor on
being handed a curve of metacentres, was only
equalled by that of a rustic on beholding the hiero-
glyphics on Cleopatra's Needle, and being told it was
a page of a book.

Until we require our nautical assessors to be men of
good technical knowledge, it is hopeless to expect that
marine courts of inquiry will ever do anything towards
solving the problem of how to lessen loss of life and
property at sea.

The skin plating in the earlier petroleum steamers
was a frequent source of trouble, especially in the
converted steamers. This is not surprising, considering that the usual method of shipyard construction
does not ensure watertightness, even after the caulker
has finished his work. Oxidation is the agent that
is generally relied upon to gradually prevent any
"weeping" at the butts and lansings. In new
steamers, unless of the very highest class, a close
inspection of the floors and frames after a few
months' service, will generally disclose the presence
of a few signs of corrosion due to weeping, which,
while harmless enough in most vessels, are very
objectionable in the case of petroleum steamers. In
one or two of the " converted " vessels, which by the
way were extremely strong and well-constructed
cargo boats, it was found necessary, to prevent straining and consequent leakage, to put butt straps inside
and out. The latter system has the disadvantage

that it materially increases the resistance of the vessel.
The better plan, and the one universally followed, is
to overlap the plates longitudinally. This system
is regarded by many as a new departure in shipbuilding ; on the contrary, it was in vogue forty
years back. A vessel built with overlapping butts
by Messrs. Reid and Son, of Greenock, forty years
ago, is still in commission and doing good work. Like
many other ideas in shipbuilding and engineering,
which in the last few years have been resuscitated
and called modern, the system of overlapping butts
was well-known to engineers and shipbuilders of the
last generation.

Petroleum being such a mobile liquid, and one not
readily combining with other substances, the very
greatest care in the construction of the vessel has to
be used ; but even in vessels of the highest class in the
petroleum trade, some little leakage is bound to take
place, especially in those vessels carying refined oil. In
those vessels that habitually carry crude oil, a deposit
of earthy matter settles out, which has a tendency to
become glutinous ; this in time acts beneficially in
stopping slight leakage.

As petroleum vessels are now constructed, they are,
without doubt, the strongest vessels in the mercantile
marine ; the numerous transverse bulkheads, with the
longitudinal middle one and the numerous arrangements of stringers and beams to prevent the slightest
amount of straining, render them, from a structural
point of view, the safest vessels afloat.

The objection is often urged by shipowners that a petroleum vessel cannot be used for any other cargo. This is a great mistake. A properly designed petroleum steamer can be most usefully employed in ordinary trades. Of course all traces of petroleum would have to be eliminated. This is best done, in the first place, by careful scrubbing down of the holds, beams &c., and using a jet of water under pressure, and finally a jet of high-pressure steam. If all traces of oil are not then dissipated, a coat of lime will effectually render the tanks sweet. Of course it is not supposed that the vessel will be employed in the tea trade or coffee trade ; but if used for coals or wood or ore for the first voyage, and then cleaned out thoroughly, the vessel could with safety carry sugar, coffee, and fine goods. All the same time a petroleum vessel would not be such an economical one as an ordinary "tramp." A petroleum vessel would not carry as much as a tramp of the same displacement by 15 to 20 per cent.

CHAPTER IV.

PUMPS. LOADING. DISCHARGING. BALLAST-TANKS. VENTILATION.

98. FOR transferring oil from storage tanks to the ship, or *vice versâ*, we require a pump that will satisfy the following conditions : the pump must be capable of giving a continuous discharge. It must be capable of working against a considerable head ; it must be economical in consumption of steam, and have a high efficiency. The working parts must be few ; it must be self-contained, and occupy little space. It must require the minimum of skilled attention, and must be of the most solid and substantial description. The pumps which fulfil these conditions are the Worthington and Tangye Duplex.

We will give a slight description of these two pumps.
In ordinary pumping plant one is met with a multi-
tude of cranks, levers, fly-wheels, eccentrics, &c., all
of which require skilled attention.   For ship use we
require pumps to work under the most disadvan-
tageous circumstances, and in the hands very often
of rather ignorant people.

99. The Worthington pump is characterised by
great simplicity and fewness of moving parts, and,
being of the most substantial construction, it stands
an enormous amount of hard work.   The beautiful
principle, introduced by Mr. Worthington, of making
the piston-rod of one cylinder actuate the slide valve
of the opposite one, commands the admiration of all
interested in mechanical movements.   In the early
days of the petroleum industry, ordinary pumps were
found to be practically useless for the purpose, and
the evolution of the Worthington pump is one of the
best examples the writer knows of, of difficulties and
obstacles overcome by patience and mechanical skill.
Such is the high efficiency attained by these pumps
when on the compound principle, that for large
pumping installations, such as waterworks, they are
superseding the well-known overhead beam engine.
The Worthington steam pump was introduced in
1844, and was the first of that numerous class of
direct-acting non-rotative pumps, whose steam valves
were controlled by mechanism more or less complex,
and which had no positive mechanical connection
between the movements of the piston and steam-

valve. Successive improvements were introduced till, in its present form, it is perhaps the simplest, most durable, and most economical pump on the market.

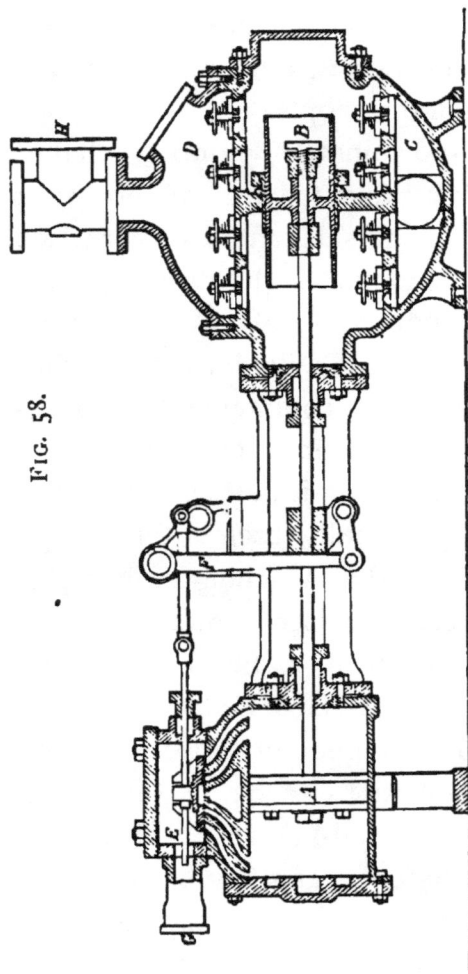

FIG. 58.

Fig. 58 is a sectional view of one side or half of a Worthington high-pressure steam pump, of ordinary construction. Its object is to exhibit the great simplicity of its interior arrangement, especially that pertaining to the steam-valve.

This valve, as may be seen at E, is an ordinary slide valve, working upon a flat face over ports or openings. Its simplicity and durability, in contrast with any other form of steam-valve, are well known. Although numerous attempts have been made to supersede it, it still maintains its place on locomotives and other forms of high-pressure crank engines. No matter how long the engine may stand inactive, a slide-valve will not rust or adhere to its seat, and is always ready to start when required. No water can collect in its cavities to produce trouble by freezing. In a word, it may be called the simplest and most reliable steam-valve known to engineers.

In the Worthington engine the motion of this valve is produced by a vibrating arm, seen at F, which swings through the whole length of the stroke, with long and easy leverage. As the moving parts are always in contact, the blow inseparable from the tappet system is avoided. Even the motion of the well-known eccentric upon crank engines is not comparable to this for moderate friction and durability.

The valve motion is the prominent and important peculiarity of this pump, as being that to which it owes its complete exemption from noise or concussive action. Two steam pumps are placed side by side, and so combined as to act reciprocally upon the steam-valves of each other. The one piston acts to give steam to the other, after which it finishes its own stroke, and waits for its valve to be acted upon before it can renew its motion. This *pause* allows all the water-valves to seat quietly, and removes everything like harshness of motion.

As one or the other of the steam valves must be always open, there can be no *dead point*. The pump is therefore always ready to start when steam is admitted, and is managed by the simple opening and shutting of the throttle valve.

Attention is directed to the arrangement of the double-acting plunger shown at B. It works through a deep metallic packing-ring, bored to an accurate fit, being neither elastic nor adjustable. Both the ring and the plunger can be quickly taken out, and either refitted or exchanged for new ones at small cost, and if it be desired at any time to change the proportions between the steam pistons and pumps, a plunger of somewhat larger size, or decreased to any smaller diameter, can be readily substituted. As exact proportions between the power and work are always desirable, if not necessary, this is a very important advantage.

This system of renewal of the working parts has been proved by long experience to be the least expensive and most satisfactory for ordinary work. The plunger is located some inches above the suction valves to form a subsiding chamber, into which any foreign substance may fall below the wearing surfaces

M

This enables it to work longer without renewal than the usual form of piston-pump, especially in water containing grit or other solid material. The water enters the pump from the suction chamber, C, through the suction valves, then passes partly around and partly by the end of the plunger, through the force valves, nearly in a straight course, into the delivery chamber, D, thus traversing a very direct and ample water way. The bottom and top plates furnish a large area for the accommodation of the valves. These consist of several small discs of rubber, or other suitable material, easy to examine and inexpensive to replace.

100. Two steam cylinder sand two pumps are cast together to form one machine. The right-hand division moves the steam-valve of the left-hand one, and *vice versâ*. Under this arrangement one pump takes up the motion when the other is about to lay it down, thus keeping up a uniform delivery without pulsation or noise. As the work is divided between two engines, the wear is also divided, and the lifetime of the machine doubled.

Added to its durability, the smooth and noiseless action of the Worthington steam-pump makes it preferable on many important services where the jar of a single-cylinder pump would be objectionable or destructive.

101. Sizes and particulars of the Worthington low service pump. Those marked thus (*) are, in the writer's opinion, most suitable for oil-steamers.

Diameter of steam Cylinders in inches.	Diameter of Water Plungers in inches.	Length of Stroke in inches.	Displacement in Gallons per Stroke of one Plunger.	Proper Strokes per minute of one Plunger.	Gallons delivered per minute by both Plungers at stated number of Strokes.	Sizes of Pipes for short lengths, to be increased as length increases.			
						Steam Pipe.	Exhaust Pipe.	Suction Pipe.	Discharge Pipe.
4½	3¾	4	·183	75 to 150	25 to 55	½	¾	2½	1½
6	5¾	6	·56	75 to 125	85 to 140	1	1½	4	2½
7½	6	10	1·01	50 to 100	100 to 200	1½	2	4	3
7½	7	10	1·38	50 to 100	140 to 280	1½	2	5	4
*9	8½	10	2·03	50 to 100	200 to 400	1½	2	6	5
7½	10¼	10	2·96	50 to 100	300 to 600	1½	2	7	6
10	10¼	10	2·96	50 to 100	300 to 600	2	2½	7	6
9	12	10	4·06	50 to 100	410 to 820	1½	2	8	7
*12	12	10	4·06	50 to 100	410 to 820	2	2½	8	7
7½	14	10	5·53	50 to 100	560 to 1120	1½	2	10	8
12	14	10	5·53	50 to 100	560 to 1120	2	2½	10	8
14	14	10	5·53	50 to 100	560 to 1120	2½	3	10	8
10	15	15	9·52	50 to 90	960 to 1720	2	2½	12	10
12	15	15	9·52	50 to 90	960 to 1720	2	2½	12	10
14	15	15	9·52	50 to 90	960 to 1720	2½	3	12	10
14	17	15	12·25	50 to 90	1230 to 2210				

102. For petroleum-steamers the author thinks
that the Worthington side pipe and strainer, Fig. 59,
should be fitted. It frequently happens, when carrying
cargoes of crude oil, that there is a considerable

FIG. 59.

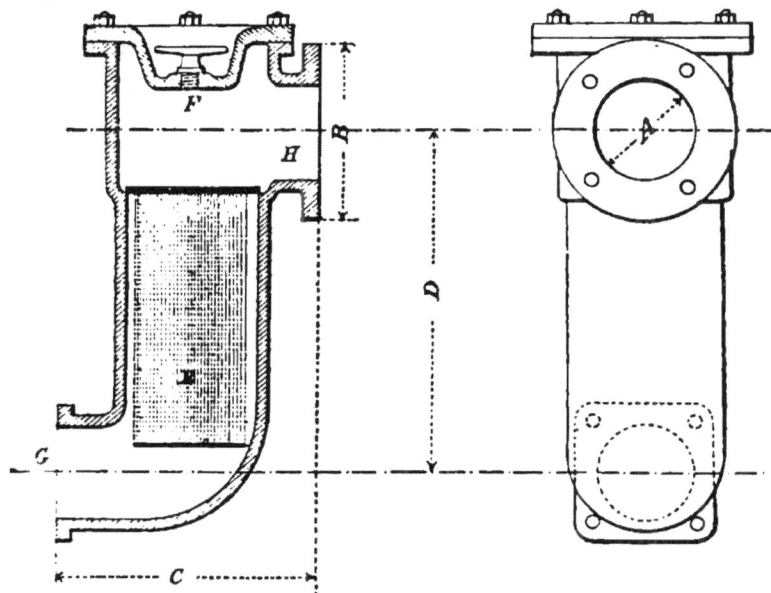

earthy sediment which deposits out in the pipe bends
and oil chambers. In the tanks this sediment collects,
and ultimately assumes a gummy consistence which
might cause trouble. By fitting the Worthington side
pipe and strainer these difficulties will not occur;
above is a sketch of the attachment.

The flange G is attached directly to the suction
opening of the pump, and the supply pipe to the

flange H. A basket strainer of ample area is inserted at E, down through which the fluid passes, and which can be easily withdrawn for cleaning through the hand-hole at the top. In the hand-hole plate a thumb-plug F is provided, which, if removed when the pump is stopped, admits air into the suction pipe, and prevents the oil from syphoning out of the oil cylinders. The pump being thus charged with oil is enabled to lift its supply more readily when again started.

103. At times, depending upon the "lift," some trouble will be experienced in making the pump work satisfactorily. Frequently the oil has to be forced some distance to the reservoirs by the ship's pumps.

Unless the suction lift and length of supply pipe are moderate, a foot valve, a charging connection, and a vacuum chamber are very desirable, if not absolutely necessary. The suction pipe must of course be entirely free from air leakage.

It often happens that a pump refuses to lift oil or water while the full pressure against which it is expected to work is resting upon the force valve, for the reason that the air within the pump chamber is not dislodged, but only compressed by the motion of the plunger. It is well, therefore, to arrange for running without pressure until the air is expelled, and oil or water follows. This is done by placing a check-valve in the delivery pipe, and providing a waste delivery to be closed after the pump has caught water. Such a valve is also required for keeping back the pressure when the pump is opened for examination of the valves.

Fig. 60 is given for the purpose of showing those

FIG. 60.

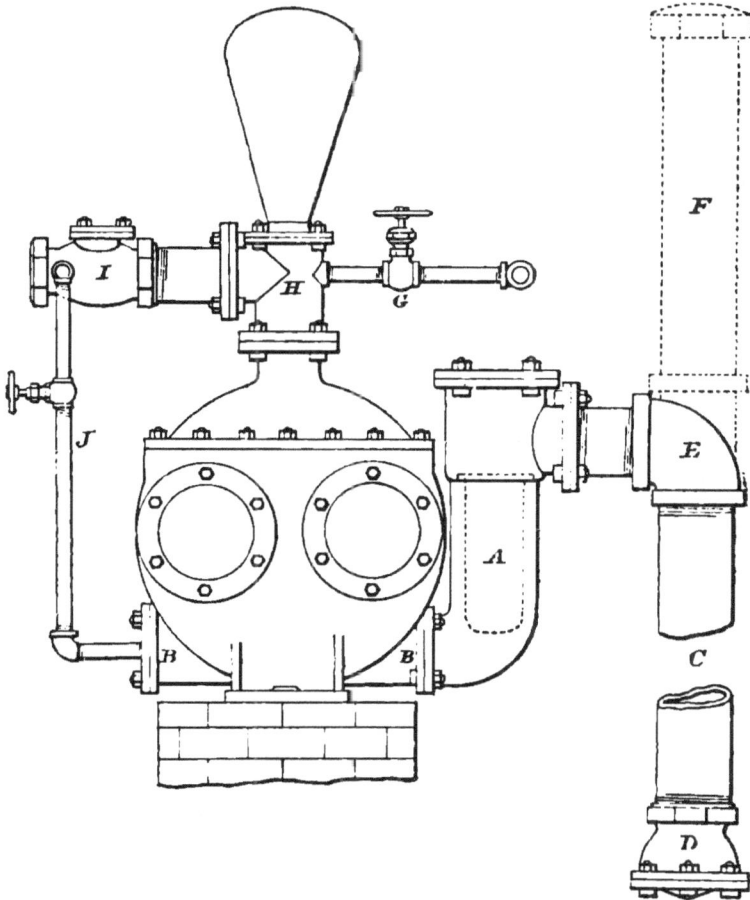

not fully acquainted with the subject a good arrangement of piping, together with the attachments above referred to.

On the suction pipe C is a foot valve D, which keeps the pipe and cylinders charged with oil, so that the pump, when being started, does not have to free itself and the suction pipe of air. This valve is always essential on an unusually long suction pipe, or where the suction lift is severe. In such cases the vacuum chamber F should also be added. It is readily made by extending the suction pipe upward, using a tee instead of the elbow E, and putting a cap on the top.

The arrangement of side pipe and strainer, shown at A, see Fig. 59, is described on pages 164–5.

A check-valve I, should be placed upon the delivery pipe, to keep back the oil when the pump is opened for examination or repairs.

A "waste-delivery" or "starting pipe," that can be led into any convenient place of overflow, should be provided as shown at G, so that the pump at starting can free itself of air, while the pressure is kept from it by the check-valve I. When the pump has properly started, the valve in this "waste-delivery" should be closed.

The "charging pipe" J, connecting the delivery pipe beyond the check-valve with the suction chamber of the pump, is for the purpose of charging the cylinders and suction pipe before starting with oil from the delivery pipe, in case they have been purposely emptied, or the oil has leaked out through the foot-valve.

The suction or supply pipe can be attached to

either or both sides of the suction chamber at the flange B. On some of the larger sizes the suction opening is on the end of the pump.

104. Inasmuch as high pressures of steam are almost universal on board ship, it seems advisable in the largest class of petroleum-steamers to introduce the " Compound Worthington " ; of course this would mean extra expense in the first place, and it would also involve either the use of the main condenser and circulating pump, or else the attachment of a separate condenser, the air and circulating pumps being worked off the L. P. piston rod, which would be lengthened. Such an arrangement would be economical, but would require a larger pump-room and perhaps an extra engineer.

105. The Tangye Duplex Pump.—This pump is similar in many respects to the Worthington, and, like it, is distinguished for simplicity, rigidity, and fewness of working parts, together with a very high efficiency. Like all the work turned out by this well-known firm of engineers, it is distinguished by its high-class finish and splendid fitting. The duplex pump is perfectly noiseless in its action.

For rapid discharging of petroleum-steamers, for feed, bilge, fire, and other purposes, the Tangye duplex is largely used and highly appreciated by sea-going engineers.

The accompanying illustration represents the type of duplex pump (Tangyes' Patent) being made in numerous sizes by Tangyes, Limited, of Birmingham, and supplied for use on petroleum-steamers. The size from which it is taken is capable of delivering 100 tons per hour. The steam cylinders are suitable for pressures up to 150 lbs. per square inch, and are 9 inches diameter, pumps 8½ inches diameter, and the stroke 12 inches each. Steam cylinder is fitted with bye-pass valves for regulating and obtaining the full

FIG. 61.

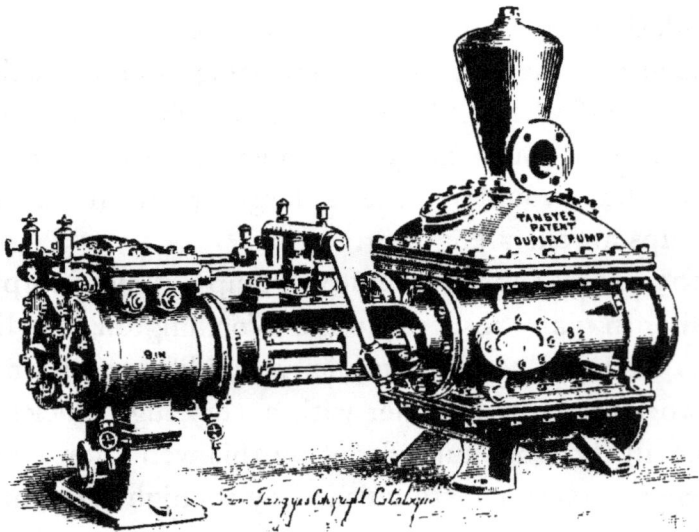

stroke. The pistons are of cast iron, fitted with metallic rings and Muntz's metal rods, the pump ends being secured to plungers by deep gun-metal lock nuts. Each slide valve is worked by a lever worked from the opposite pump rod, and a certain amount of slip is allowed, to obtain a pause at each end of the stroke, so as to ensure the pump valves returning to their seats gently. There are no pins in the valve motion to wear or to get loose. The swivel joints are bushed with gun metal, and fitted with oil caps, so that the ends of levers shall always work in lubricant. The pumps are double-acting, fitted with gun-metal liners and rams. The liners can easily be renewed, as the securing bolts are kept outside pump. The suction and delivery valves with their seats are of gun metal with manganese bronze spindles screwed and riveted in.

These valves are carefully secured in two cast-iron valve plates, one for suction, the other for delivery, which being interchangeable do away with the necessity for taking more than one spare plate to sea. Extra large and convenient hand holes with covers and air vessel is supplied, and the steam cylinders are fitted with connecting steam pipe, condensed water cocks, and lubricator.

106. Pump-room.—This important part of the ship should be situated midships, and if possible should be on the lower deck, as the nearer the pump is to its work the better. It should be large, airy, and well lighted. The pumps should be placed so that they are easily accessible for examination ; the main fitting and suction pipes should be so arranged that all the connections are as direct as possible. Bends and elbows should be dispensed with as much as possible ; in fact the skill of the designer is very effectively shown in the arrangement of the piping. Simplicity is the great thing. Pipes and valves

should be numbered and coloured for their respective tanks, e.g. :—

"No. 5 after starboard tank" should be painted green, with "No. 5" in bold white letters. A plan showing the lead and dimensions of all pipes and valves should be provided by the builders and hung up in a conspicuous place. While loading and discharging, the pump man should never leave the pump-room. Indicators should be provided showing the state of each tank.

107. Usually, too, in a pump-room all the filling and discharging pipes will enter either on a vertical trunkway, or else be led in anyhow, along the deck &c. This latter practice has nothing to recommend it, and is only evidence of bad design and want of thought. In any case means should be taken to see that each pipe can be traced and examined for any leakage. All the pipes leading to the tanks should be easily accessible, as it is possible that in very cold weather a deposit of paraffin might cause an obstruction. Steam pipes from the main and donkey boilers, for the oil-pumps, should not be led along the upper deck, as is usually the case. They should be led along under the upper-deck and suspended in hangers from the beams and carefully jacketed.

If the pump-room is below decks a speaking-tube or electric trembler should be provided to enable the officer in charge to signal to the pump man. A place should be provided for the storage of connecting pipes, spare valves, nuts, bolts, &c., and there should be a good supply of spanners, cold chisels and hammers. A vice bench should be erected, and everything, in fact, supplied to enable work to be efficiently and expeditiously carried on. It should be a stringent rule that nothing should be kept in the pump-room that does not properly belong to it. It should not be used as a bo'sun's locker, or paint store.

108. Pipes.—The vessel being moored alongside, or stern on to the quay, the pipes connecting the delivery valve on the quay with that on board are connected up. In some cases these pipes are made of Indiarubber or canvas, with a spiral wire inside. They are made in lengths of 5 to 7 feet. The ends are opened and the flanged socket introduced, G bands (2 or 3) of wrought-iron are then passed on, and the ends set up with a bolt and nut. Pipes made on this principle do not, however, last long, and if the pressure in the mains is anything more than moderate, and especially if there is any knock in the pump, they give out very soon. If these indiarubber pipes are used they should not be allowed to hang in a bight, but should be supported by strops and light tackles, taking care that the strop is put round the pipe at the junction. If there is any knock in the main it will be advisable to keep the pipe suspended, as the otherwise continued jerk will soon make it self-evident.

109. The life of indiarubber pipes can be much prolonged by covering them with canvas and "serving" over with well-stretched ¾-inch Manilla rope

In joining up the flexible pipes do not be afraid to

use plenty of bolts. Usually there are five holes in the flanges; a joint can be, and very often is, made with merely a couple of bolts; it is better to use four or five, as the pipe is then equally strained in all directions. Iron bolts with solid heads and hexagon nuts are not expensive; bolts (with nuts) $\frac{3}{4}$-inch diameter and $3\frac{1}{2}$ inches long cost about 25s. per cwt.

110. Much better results are obtained by the use of wrought-iron pipes with gun-metal elbows and universal joints. Such pipes are made in any length with a good broad flange. A few suitable lengths should be kept already joined together.

In joining pipes it will be well to be provided with a good supply of nuts and bolts; these should be kept in a wooden box, and occasionally some hot white lead and tallow poured over them to make them easy to work, and also to keep down rust. Nothing is more vexing than to keep a gang of men hanging about while rusty bolts and nuts are being run up in a vice. Several spanners and that useful tool the " marlingspike " should be at hand.

111. As making a good joint is very often a difficult matter with new hands, we will describe in full how it is done. First, see whether the flanges of the two pipes will allow a joint to be made without the use of clamps. Flanges usually have five bolt-holes. See that the pitch circle of the bolt-holes is of the same diameter in both flanges. It can easily be ascertained whether a joint can be made, by bringing

the two flanges together, and putting the marling-spike or a small drift through two holes, seeing whether the others coincide, or whether they are "blind." Very often the holes will overlap, in which case, by turning one flange round, a position will be found in which all the holes are fair. This being so, insert two bolts in the lower holes, and put on the nuts by hand. A ring of paper is then inserted between the flanges, and the other bolts put in and

FIG. 62.

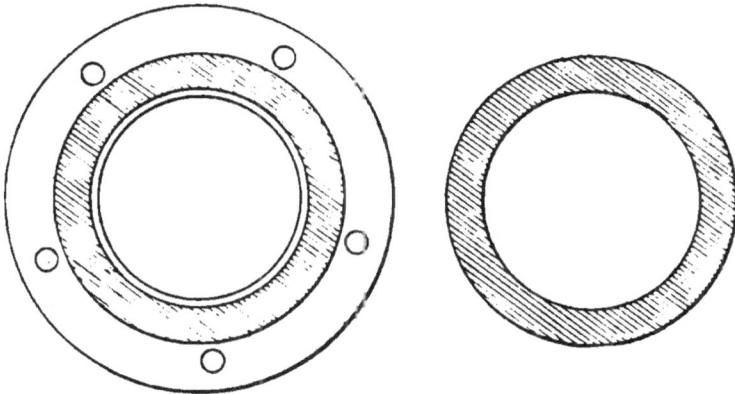

hove-up with the spanners. Stiff paper is most effective in making a good oil-tight joint, and a good supply of paper rings should always be kept in hand.

Any thick paper will answer the purpose, although millboard is of course the best. These packing rings should be cut out with a carpenter's chisel. A template of Muntz metal is very useful for marking out. The rings should, in inside diameter, be the same, or just a little larger than the diameter of the pipe; the outside diameter should be a little less than the greatest distance (diametrically) between the inner edges of the bolt-holes, Fig. 62

Before using these rings, they should be dipped into water and then (if for a permanent joint) into glue. If they are made of thin paper several should be glued together, so as to make them of the same thickness as if made out of millboard. Rings made

FIG. 63.

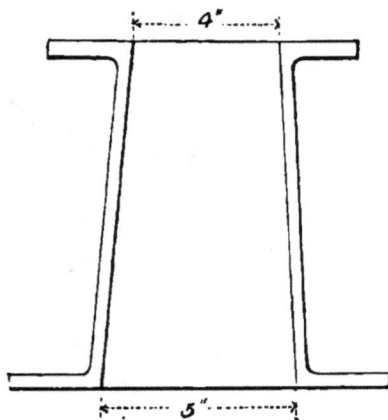

from old book covers, old charts, and brown paper answer admirably.

112. It may frequently be the case that it is necessary to join up a pipe with a larger or smaller one, in which case a small length of piping is used, Fig. 63. Should such a pipe not be at hand, we must use clamps. These clamps, see Fig. 64, merely consist of two similar pieces of metal, with a nut and bolt at each end. They are opened so as to allow them to be

slipped over the two flanges of the pipe, and then hove-up tight.   Two pairs will be sufficient to join two pipes together, but the use of three makes a better job.

113. If gun-metal universal joints are used, great care must be taken to arrange them so that all possible motions of the ship are provided for.   Three joints are requisite and two lengths of piping.   The joint

FIG. 64.

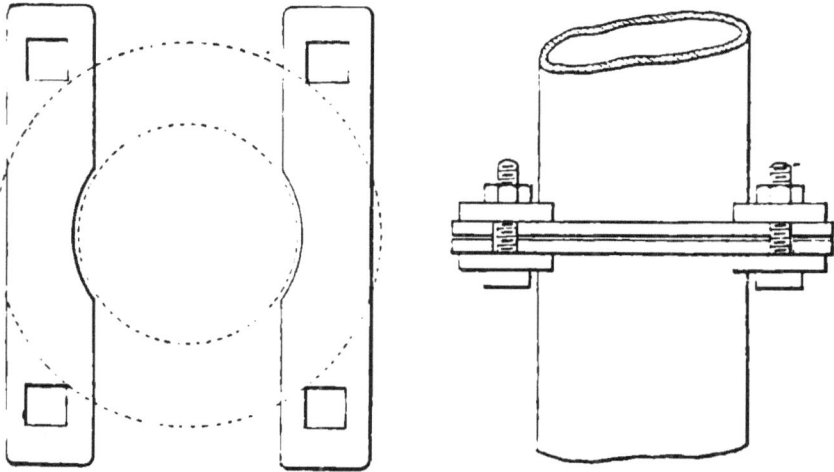

should be placed between the ship and the first end of the pipe, one between the two lengths of the piping, and the other connecting the other end of the piping with the delivery valve on the quay.   This arrangement allows the pipes to lay in a V, and all motions are provided for.   It will be advisable to hang the pipe up by means of a strap and tackle, so as to take the weight and prevent any stress being set up in the elbows of the universal joints. These latter should have plenty of play. Common gasket packing does very well for insuring tight joints. When loading and discharging, leakage is a sign of carelessness.

114. Loading.—All the necessary connections being made, the officer in charge of the pump-room should open the stop-valve on the pipe leading to that tank which is to be filled first. See that the air in the tank has some means of exit, either through a manhole in the tank top or else by a ventilator. See that all the valves on the other filling pipes are shut; also the valves leading to and from the pumps are closed. The delivery valve on the wharf can now be opened, and the oil admitted into the tank.

In filling tanks the ones at the ends of the ship should be filled first, as such tanks being narrow and small there is not such a loss of stability when they are half filled ; and when completely filled, they of course increase materially the stability of the ship ; on the other hand, by filling the end tanks first, great stresses are brought upon the ends of the vessel. Many people think it better to fill the midship tanks first and empty them last. Every case will depend upon the construction and dimensions of the ship.

115. It may often happen that in filling one tank, or filling two wing tanks simultaneously, the vessel will gradually heel over, especially if the other tanks are empty. It depends entirely upon the amount and range of the stability of the vessel, whether such heeling need cause any apprehension.

There are some vessels carrying oil in bulk that have very little initial stability when all tanks are empty and ballast tanks full, and in filling the tanks they gradually take a strong list, generally to port. If now the officer in charge opens a valve in one of the starboard wing tanks, and puts a few tons of oil in, he will find that the vessel will gradually trim upright, and to his disgust will take a heavy list to starboard. If a vessel lists easily in the early stages of loading, the only thing to do is this: fill the tank right up. This will give the vessel more stability, and by the time one or two tanks are quite full, the vessel should be perfectly upright and fairly stiff.

Do not be too anxious to correct a slight list by putting a few tons of oil into the wing or trimming tanks. If you start on one tank do not leave off till you have filled it.

116. Having nearly filled one tank, the pump man should stand by to change the valves. There should be an efficient indicating apparatus in the pump-room which shows the condition of each tank. As soon as the expansion chamber of the tank is as full as requisite, the valve on the filling pipe of the next tank should be opened, and by manipulating the valves in an intelligent manner the tanks can be filled without stopping the pump on the quay.

Recollect NEVER *shut one valve without opening another, and always fill each tank full as possible.*

When filling the last tank it is advisable to cease pumping before the tank is quite full, and allow the oil in the pipe to drain into the tank ; or, better still, an air pump is put in operation, and the remaining oil in the pipe blown through. If necessary, one or two extra strokes of the pump can be given to complete the filling. While loading and discharging, great care should be taken that no naked lights are used on any pretence.

117. Discharging.—This is done with the vessel's own pumps, but steam is supplied in many cases from the quay. The foregoing remarks as to pipes and joints apply in discharging. Pumping seems a very simple thing, but it is easily seen that, like everything else, it requires care and skill; any one can turn a valve, but in draining a tank considerable *nous* is requisite. In the large tank-steamers now built a great distance exists between the pumps and the tanks in the ends, and it will be found that while a too slow stroke will not lift the oil, neither will a too rapid one ; but it is quite possible to drain a tank and empty the "sumpt" if the pumps are driven in an intelligent manner; and captains are advised not to allow a quantity of loose oil to wash about in the bottom of a tank simply because the excuse is made that the "pumps won't suck." The difficulty is in keeping the pipes charged with a column of oil : check the steam-valve, and if there is another full tank of oil, just open the valve a bit and let the pump suck both from the empty and full tank ; by this means the vacuum can generally be re-established. As soon as the pump takes the oil the valve on the full tank

can be closed. In some vessels a steam pipe is fitted from the main steam pipe to the main suction pipe of the pump; the object is, by allowing a jet of steam to enter the suction-pipe the steam is condensed against the cold oil, and a vacuum is of course made. In general this "dodge" should not be used.

In the case of crude oil, steam being injected into the suction simply causes the generation of vapour, which is fatal to the establishment of a vacuum ; and in the case of refined oil, the condensed steam, falling in the form of water, produces cloudiness in the oil, which is fatal to its sale. The same rule holds good in discharging as in loading, and that is—*always empty one tank before commencing on another.*

118. It may happen that the vessel has discharged say a cargo of crude oil, and is then to take in a cargo of refined oil. Any change of colour, or cloudiness, in the latter seriously affects its market value. It will be necessary to carefully wash out the tanks, and clear the pipes from all remains of crude oil. The best plan seems to be that of pumping in a few tons of sea water, and carefully brooming down the sides and tops of the tanks. A hose with a nozzle and a good head of water is very effective in removing the earthy deposit. After cleaning all the tanks thoroughly, the water, oil, and muck is pumped through the pipes overboard. A few barrels of refined oil may then be used to give a final wash down. Stages may be rigged on the tanks, and the deck hands provided with hand scrubbers, and a few buckets of refined oil

can give all surfaces a good wipe down so as to remove all particles of water and muck. This is carefully baled out of the sumpt. Should there be any doubt as to the cleanness of the pumps and pipes, steam can be injected into them, and the valves opened one by one. All the dirt will thus be deposited in the sumpt. The dirt and the condensed water can then be carefully taken out. Generally the pipe leading into the sumpt is used both as a filling and discharge pipe. It should be fitted with a rose which should be easily detachable.

Another way to clean tanks is to play a jet of high-pressure steam or hot water against the sides and tops of the tanks. This is very effective, but heating the plates results, which causes the exfoliation of the outside paint, and assists in setting-up rust cones. The expansion and contraction of the plates may cause leaks.

119. When examining and cleaning oil tanks, there should always be some one stationed on deck to give an alarm. The fumes of petroleum produce on many people the effects of intoxication, similar to those produced by remaining long in a spirit vault ; and one or two accidents have occurred by sending a man into a tank without keeping a watch to see he does not be overcome by the vapour.

120. In many vessels it is the practice after the tank is empty to close it up, leaving one or two openings in the ventilators, and inject a quantity of high-pressure steam ; this has the effect of pulverising the petroleum vapour, and it is carried up through the

ventilation into the air. This is a very good method
of cleansing the tank from vapour. To inject the
steam, it is as well to have a number of lengths of
3-inch iron pipes with screwed sockets, and one or two
flexible joints.

121. The oil tanks should be provided with easy
means of access. The method adopted in some
vessels, of having the lids of the tanks of large area
and secured by numberless nuts and bolts, is not to
be commended. The better method is that adopted
by Mr. Swan, of Messrs. Armstrong, Mitchell and Co.,
by which the lid of the tank can be removed in one or
two minutes, merely by turning a few screw handles.

122. Ballast tanks.—In vessels fitted to carry water
ballast, unless great care is used, ballast tanks of the
ordinary construction are a constant source of vexation
of spirit, and even danger. The usual method of construction, as used in ordinary cargo vessels, is not
applicable to oil-carrying steamers. In the latter it
is necessary that the tops of the ballast tanks should
be of the same construction, so far as oil-tightness is
concerned, as the hull of the ship ; and if the ship has
the slightest tendency to strain, the tops of the ballast
tanks will invariably give evidence of the fact, and
leakage will result. In some vessels of the "converted" type, much trouble has been caused through
leakage in consequence of bad design in this matter.
Those who have followed our remarks on stability will
readily see that an accumulation of a few tons of
loose oil in the ballast tank is in itself a serious

element of danger, more especially when it is recollected that in these "converted" vessels, so far as the writer is aware, no connection is made, as there should be, between the ballast tank and the oil pumps. Consequently, if leakage occurs, it has to be got under by pumping through the ballast pump in the main engine-room, a most objectionable proceeding, especially if crude oil is carried, as thereby one of the canons of the safe carriage of petroleum is violated, which is that *under no circumstances should petroleum, forming the cargo, be allowed to enter the boiler or engine room.* In addition to the imprudence of using the ballast pump and the engine-room valves for petroleum, the oil thus pumped is pumped overboard, thus entailing loss which has to be paid for.

123. If the ship is fitted with ballast tanks, the ballast pump should be placed in the pump-room, and if the piping and valves are arranged properly, there is no reason why the oil pump should not serve both purposes.

Connection should be made between the ballast tank and the oil tank, so as to allow any leakage to be pumped back again. A better plan is to let any leakage run into a separate compartment, so as to allow it to settle, as, in carrying refined oil, any admixture of water would cause cloudiness.

124. If through leakage any one of the oil tanks ceases to be quite full, and bad weather is encountered, it would seem prudent to fill it up at once with water, and if there is a quantity of loose oil rolling about in

the ballast tanks the safest course is to pump it overboard. Sometimes the ballast tank is allowed to get
quite full of oil; and, failing any better means of
dealing with it, it is usual, when the oil tank over it is
pumped out, to take off the manhole door in the ballast
tank and open the sea cock of the ballast tank, the oil
is then forced through the manhole and runs away to
the sumpt. This is a barbarous method, and also most
objectionable. Looking to the difficulty of making
a ship's tank quite impervious to oil, the writer is of
opinion that ballast tanks are not desirable in petroleum vessels—in point of fact, they are not at all necessary, as there is no reason why one or more of the oil
tanks should not, on the return voyage, be filled with
water. This is the case with such well-designed
vessels as the *Oka, Era, Charlois, Lux,* and others.

Should a leak occur in a tank at sea, it will depend
upon the rate of leakage, the duration of the voyage,
season of the year, and the quality of the oil, as to
what measures should be taken. If the cargo in this
tank consists of crude oil, the addition of water will
not greatly matter, and indeed this is often the only
course to be followed. If, on the other hand, the cargo
is refined oil, and the ship is but a few days from her
destination, the weather is likely to remain fine, and
everything in fact being favourable, it would seem to
be advisable not to refill the tank unless actually
compelled to, because the addition of water or an
inferior quality of oil will simply result in the con-
signees refusing to take delivery. If, however, the

captain is certain that the loss of stability occasioned through having a tank partially filled is likely to imperil his ship, he must of course either have the tank empty or full.   It may be possible to fill a large tank from a smaller one or *vice versâ*, and so make up the deficiency.   However, such a question must be decided in the light of surrounding circumstances, and it is not possible to give directions for every case. The reader, bearing in mind our remarks upon stability, will see the decision of the captain, in the event of leakage, must be guided by many circumstances.

125. While on the subject of leakage, a few words may be said as to the methods of stopping a leak. There are many patent compositions in the market which are supposed to make a leaky seam oil-tight. So far as the writer is aware, the only cure for leaky tanks is to have thorough good workmanship—in fact, tanks should be a boilermaker's job.

Portland cement can be used in certain circumstances.   To obtain good results it is necessary that the structure should be perfectly rigid in the vicinity of the leak, the cement should be perfectly anhydrous and new, and the surfaces of the metal should be freed from every trace of oil.   If a leak is situated in a corner, Portland cement can be used with advantage. All the surrounding metal should be washed with a strong solution of caustic soda, and all the rust removed ; the surfaces should be roughed with a sharp cold chisel, so as to make sure the cement will have something to hang to.   The cement should be mixed

with about one-third of clean sharp river-sand, and if
desired to set very quickly a little sugar or molasses
should be added, in the proportion of about 2 lb. of
sugar to each bushel of cement; the cement is then
applied. To ensure good results this should be done
in port, as any vibration will prevent the cement
adhering. If carefully done, this method of dealing
with a leak is very effectual. Caulking is of doubtful
advantage. If the workmanship on the tanks is
good and the tanks themselves very substantial in
construction, it will give good results, but not
otherwise.

126. Manholes.—Vessels that have a double bottom,
in which the top of the ballast tank forms the bottom
of the petroleum tank, generally have manholes fitted
for purposes of access to the ballast tanks; unless
extreme care is taken in fitting these manholes, they
remain a source of constant trouble through the
leakage that takes place from the oil tank to the
ballast tank.

This is especially the case with those vessels that
have been converted. The very greatest care is
necessary in fitting the manhole doors.

Compensation rings, inside and out, should be
riveted to the tank top. The door should be of cast
steel. The dogs of very substantial make. Rings of
thick paper make the best packing. Inasmuch as
leakage from the ballast tank to the oil tank is not so
important as the reverse, the manhole door should
butt on the upper side of the ballast tank top, and the

weight of the superincumbent oil will tend to further ensure tightness.

127. Sounding pipes and air pipes should of course be fitted. It frequently happens that during the filling of a ballast tank, a column of water, or oil if it is there, is injected up through the air-pipe and discharged over the deck; this is a clumsy method which should not be permitted. When the screw cap of the air-pipe is removed a valve should be screwed in and opened to allow the air to be expelled. When the water or oil makes its appearance, a short length of hose let into the scuppers will transfer any oil or water overboard. This should be attended to in the case of vessels which have wooden upper decks, as these would in time become saturated with oil and might prove a source of danger.

The writer once saw a captain of an oil tank vessel very carefully assisting his crew in sweeping a quantity of crude petroleum over his wooden decks, and was very proud of his idea of preserving wood! The idea was of course good, but just imagine the dense stupidity of a man who could thus incur a terrible risk in order to preserve his decks! What wonder that, with such men holding responsible positions, explosions occur?

128. Sluices.—Connection should exist between two adjacent compartments by means of a "sluice" at the bottom of the tank, the spindle being led to the upper deck through a watertight gland. It is very important that these sluices should be oil-tight;

the exigencies of the trade might render it necessary to carry very different grades of oil in adjacent compartments, and any admixture would cause the whole to be seriously depreciated in value.

129. Oil-tight sluices should be fitted to the longitudinal middle bulkhead, and particular attention should be bestowed to see that they are kept in the best working order.

The rule in ordinary vessels is to keep the sluices open at sea. In petroleum-carrying vessels each compartment should be kept perfectly distinct, and the sluices only opened as required. As regards the drainage of the ship the practice usually is to allow everything to run into the main bilge pump, from whence it is discharged by the pump worked by the main engines. It seems advisable in petroleum vessels that this practice should not prevail, but that all leakage from the tanks should run into a chamber from whence it can be pumped back either into the tanks or else overboard. The main discharging and filling pumps should be so fitted as to not only fulfil these purposes, but also to serve for fire purposes.

130. Ventilation.—Good ventilation is very essential in a petroleum-carrying steamer. In carrying refined oil little or no disengagement of vapour takes place at ordinary temperatures, but we may soon expect to see steamships carrying oil in bulk down the Red Sea. When a tank is full, owing to the small area of oil exposed to the atmosphere, very little danger need be apprehended even when in the tropics ;

but it is always advisable to dilute the vapour, so as to render it harmless.

Loading and discharging are the most dangerous times in the career of an oil tank steamer. If the vessel is fitted with a fan for ventilating purposes, it would be advisable to let it run at these times ; but the writer's own opinion is that fans are not by any means a necessity, and are generally a source of trouble and tribulation to the engineers, besides taking a deal of steam to drive them.

Indeed the conditions on board ordinary vessels are not by any means favourable to high-speed machinery. Marine engineers have a rooted and very natural objection to anything which revolves above the normal speed of marine engines, and fans, which make from 200 to 700 revolutions per minute, are a source of vexation of spirit, besides being very liable to give out. It is too much to expect the usual engine-room staff to keep in order a set of pumping and ventilating engines; these latter require very careful attention, and hence the writer does not advocate their use in ordinary vessels. If they are used they are best driven by an electro-motor.

131. Ordinary canvas windsails are very good ventilators, but cannot be carried in all weathers. When in ballast a small quantity of oil in the tanks will disengage considerable quantities of vapour, especially if there is much motion on the ship ; it is very desirable that this should be got rid of as fast as possible.

The plan the writer recommends is to employ Boyle's ventilators, which are extremely efficient.

Avoid the ordinary ship's pattern ventilators—those abominations that are made of $\frac{3}{16}$-inch iron, and which have done more as a cause of foundering than many people imagine. These things are unable to withstand any pressure of water, and are liable to be lifted up by a sea and swept overboard. Some vessels have an elaborate system of ventilation, which on a breakdown of the fan is rendered useless. The simple means of taking advantage of every opportunity to remove the tank lids and hatches seems to be as good as any. Cabins and living-rooms should have efficient ventilators that can be used in all weathers, as, where there is smoking, &c., an accumulation of vapour is dangerous. A good ventilator should be capable of being used in all weathers, and should be placed sufficiently high to be out of the reach of the sea. If bell-mouthed ventilators are used they should be fitted with a diaphragm of wire gauze; ventilators should not expel the contents of the tanks too near the deck. A few ventilators are better than many, provided proper attention is given them; special precautions should be taken when loading, as in most cases the ventilators act as air-pipes, and the vapour, being rapidly disengaged as the tank fills, it rushes up and is discharged all over the ship.

132. After a petroleum tank is emptied, the oil, especially if of heavy specific gravity, clings to the sides, and we then have a small amount of oil distributed

over a very large area ; and if the temperature of the tank is what is termed temperate, or anything above, we have all the conditions necessary for the production of a dangerous vapour.

As we have before said, one of the products of petroleum is methane, or marsh-gas. If we produce this on a small scale, and fill a jar with it, and plunge a lighted taper into it, taking care that no air enters, the taper will be extinguished ; if the mouth of the jar is in full communication with the air, we shall, on application of the lighted taper, produce an explosion, and the principal gas produced by the explosion would be carbonic dioxide, or, as it is incorrectly termed, carbonic acid. We might conduct a series of experiments by means of which we should determine the amounts of air and gas necessary to produce explosion, and we might also determine the amount of air necessary to dilute the methane so as to render it harmless.

133. Experiments have been carried out in this way, and since the explosion of the *Ville de Calais*, experiments have been made to determine what proportion of air must be present to form an explosive mixture. With 1 of petroleum vapour to 5 of air no explosion occurs, but when the air is to the vapour as 6 to 1, the mixture is feebly explosive, and becomes violently so when there is 1 part of petroleum vapour to from 7 to 9 of air. With 12 parts of air it still explodes violently, but with 16 parts it is but feebly explosive, and with 20 parts of air it will not usually explode at all.

Colonel Majendie has also carried out some experiments with regard to the more volatile products of petroleum, and he concludes that one volume of liquid benzine will render 16,000 volumes of air inflammable and 5000 volumes violently explosive. Though these results show that great care is necessary in storing benzines and crude petroleums, other of his experiments are more reassuring, as he has found that neither a glowing coal, sparks from a flint or steel, or a flameless fusee, will ignite the most explosive mixture of petroleum vapour and air, actual contact with a flame or white-hot body being necessary.

The Russian chemists, Jawein and Lamansky, have been conducting experiments to determine the explosive properties of a mixture of naphtha gas and atmospheric air. They prove the mixture to be explosive when it consists of one volume of the gas and from 5·6 to 17·7 volumes of air—that is, when the mixture contains not less than 85 per cent. of air and not more than 94·4 per cent. It takes less of the gas of naphtha, therefore, than of fire-damp to make an explosive mixture with air.

From the foregoing it will be seen how necessary it is, before commencing any repairing or cleaning operations which may involve the use of a naked light, to see that the greatest care is used to have thorough ventilation. Of course, it is impossible on board ship to make a rough analysis of the air in an oil tank to determine whether it be safe or not.

Safety consists, in all cases, in rigidly preventing the access of flame anywhere near a tank.

134. Should it be necessary to conduct extensive repairing operations, it would seem best to thoroughly wash out and scrub down the top, sides, beams, stringers, &c., of oil tanks and remove all deposit.

135. The following data with regard to ventilation may prove useful.

It is estimated that the quantity of air vitiated per minute per adult is equal to one cubic foot, and that ten cubic feet per minute are not more than necessary to maintain the air in a wholesome condition.

The composition of the atmosphere in its normal condition of purity has been ascertained to be, in 1000, 788 nitrogen, 197 oxygen, 14 moisture, and 1 carbonic acid gas.

When saturated with moisture, the air contains		Moisture should be present in the air to the amount of little less than	
Deg. F.	Grains per cub. ft.	Deg. F.	Grains per cub. ft.
30	2	50	3
41	3	60	4
49	4	70	5 nearly.
56	5		
66	6		
70	7		

Volume of air expired by an adult per minute = 480 cubic inches, of which 4·6 per cent. is carbonic acid.

O

0·686 cubic foot carbonic acid per adult per hour = 65·86 cubic feet in six hours from sixteen adults.

Four volumes of atmospheric air × one volume carbonic acid, extinguish flame. A horse is estimated to give out $79\frac{1}{10}$ ounces carbonic acid in twenty-four hours.

					Sp. Gr.
Atmospheric air ..	..	..	..	..	1·000
Oxygen gas ..	..	..	..	..	1·106
Nitrogen gas ..	..	..	..	..	0·972
Carbonic acid gas	..	..	..	..	1·524
Carburetted hydrogen gas		..	..	..	0·558

Carbonic Acid is formed of two atoms of oxygen and one of carbon, its specific gravity is 1·524 ; it extinguishes flame, and is fatal to animals. The above gases combined *mechanically* (not *chemically*) constitute atmospheric air. Carbonic acid, called variously by the coal miners "black-damp," "stythe," "choke-damp," &c., is frequently met with in mines. It is produced in all collieries by the breath of the workmen (each man exhales 6·3 gallons of this gas hourly), the burning of the lights, the explosion of powder, the fermentation of animal and vegetable substances, &c. Air mixed with one-tenth of this gas will extinguish lights ; it is very poisonous, and when the atmosphere contains 8 per cent. or more there is danger of suffocation.

136. Cooking.—The ship's galley has been the cause of loss by fire of one petroleum vessel, and it is easy to imagine circumstances that would render this possible.

Cooking should not be done by fire on any steamers

carrying dangerous cargoes, as it simply means a wasteful and dangerous expenditure of coal.   It is in all cases more safe and economical to cook by steam. There are many forms of steam ranges suitable for ships' use.   Steam of high pressure is not required, and as vessels now use steam at any pressure between 90 and 160 lbs. on the square inch, the cooking range should be of very substantial construction.   A reducing valve and spring gauge should be fitted to the steam chamber.   The reducing valve should be out of the reach of the cook, and should be frequently examined.   All the cabins and forecastle should be heated by steam heaters.

137. Smoking is of itself not dangerous, as even in an explosive atmosphere a piece of lighted tobacco will not determine the explosion ; but as smoking involves the use of matches ordinarily, provision should be made to do without them.   Probably the best means is to have a dish of heated charcoal, which will remain incandescent, but does not burst into flame.   However, minute precautions are not, or should not be, requisite in a well-designed vessel ; and if proper ventilation is assured, ordinary care and prudence will prevent any danger.   As a matter of precaution, the crew should engage, in the articles, to cease smoking at the request of the captain, as on board ship there are always men who will not see any danger.

138. Lamps.—Oil lamps or candles of any sort should not be permitted.

139. Expansion of petroleum.—For practical purposes it may be assumed that petroleum increases in bulk by one gallon in every 200 for an increase in temperature of 10° F. For a range of temperature of 40° F. the expansion is 2 per cent., and in designing tank steamships, this amount of expansion is the least that should be provided, as when the bulk trade to India and Australia is established, it will not be uncommon for a vessel to load at Batoum and while in the Black Sea experience a temperature of 30° F., and on a passage to Bombay experience a temperature of 95° or 100° F., which is an increase of 65° F., or 3 per cent. Under such circumstances it will be well to provide a liberal space for expansion.

For the purpose of estimating the amount of petroleum in a tank, sufficiently accurate results for all practical purposes are obtained by adding or subtracting from the observed specific gravity of the oil, ·004 for every 10° F. above or below 60° F.

140. Like mercury, alcohol, water, &c., petroleum expands by heat and contracts with cold. Its rate of expansion and contraction, or, as it is termed, its " coefficient of expansion," is well known.

The coefficient of expansion of a liquid is a number that denotes the increase in volume that the liquid experiences for an increase in temperature of one degree ; thus mercury, for every degree, expands about 1·5 in every 10,000 ; water expands 4·6 in 10,000, and alcohol expands 11·6 in 10,000. Thus, suppose we had a tank containing 10,000 cubic feet of alcohol at a temperature of 15° C. = 59° F., at 16° C., or 60°·8 F., the volume would be 10,000 + 11·6 = 10,011·6 cubic feet.

141. The coefficient of crude petroleum varies according to the proportion of the more volatile hydrocarbons present, as shown by the following table.

Sp. gr. at 15° C. or 59° F.				Expansion coefficient for 1°.
Under ·700	..	..	..	·00090
·700–·750	..	..	..	·00085
·750–·815	..	..	..	·00080
·800–·815	..	..	..	·00070
Over ·815	..	..	..	·00065

(REDWOOD.)

142. Isolation of the cargo. — The efforts of designers have been directed to providing efficient division between the engine and boiler rooms and the oil tanks. By placing the machinery aft, a very economical and efficient division is obtained. The bulkhead at the fore part of the boiler-room should be perfectly watertight, and should be sufficiently well stayed to withstand the pressure of a body of water extending over its whole area. A few feet further forward should be the bulkhead that forms the side of the aftermost tank. The spaces between these bulkheads is generally filled with water. An arrangement consisting of a flexible pipe and a float should be fitted in connection with the pump, so that in case of oil leakage into this cofferdam, it can be immediately drawn off from the surface of the water, and either pumped back or overboard. The arrangement pro-

posed would be arranged so that the end of the pipe
would be about an inch above the level of the water,
so as to only suck the oil.   The flexible pipe would
be kept in this position by means of a float.

This water-space should be fitted with pipes, so
that connection can be made by a valve with the
oil pump and also the ordinary ballast pump.   The
top of this space should form a watertight flat, and
should have properly fitted manholes, and also an
iron ladder to facilitate access to the bottom.   This
water space need not necessarily be full of water,
although as a matter of prudence it is better to keep
on the safe side.   In any case it should *not* be used
as a store-room, as this would render it a distinct
source of danger.

Providing all the tanks are tight, there is no special
reason why the cofferdam should be kept full of
water.   Air is a fairly good non-conductor of heat,
and if attention is paid to ventilation the tanks next
the boiler room should not be of any higher temperature than those more remote.

CHAPTER V.

TESTING.

143. IT is evident that petroleum, comprising under that generic title such highly volatile products as naphtha and such fixed ones as astaki, will, for purposes of transport, be either a dangerous or perfectly safe cargo. And it is very necessary that those who are engaged in the petroleum trade should be conversant with some means of identifying the oil they are carrying, and also of estimating with some degree of accuracy its dangerous qualities.

144. A grade of petroleum is recognised and classed (*a*) according to its colour; (*b*) smell; (*c*) its specific gravity at a given temperature; (*d*) its flashing point; (*e*) its firing point.

145. Little need be said here as regards the first two, but the colour is determined by means of a chromometer, the invention of Mr. R. P. Wilson. This

instrument is fitted with two parallel tubes furnished with glass caps, and at the lower end of the tubes is a small mirror, by means of which light can be reflected upwards through the tubes into an eye-piece. One of the tubes is completely filled with the oil to be tested, and beneath the other tube, which remains empty, is placed a disc of stained glass of a standard colour.

On adjusting the mirror and looking into the eye-piece, the circular field is seen to be divided down the centre, each half being coloured to an extent corresponding with the tint of the oil and of the glass standard respectively. An accurate comparison of the two colours can thus be made. The glass discs—which, for the English trade, are of five shades of colour, termed "Good Merchantable," "Standard White," "Prime White," "Superfine White," and "Water White"—are issued by the Petroleum Association of London.

146. As regards smell, the test will of course vary in value. Some people have keener organs of smell than others, and indeed one requires to be well up in the subject before one's nose can be relied upon. A trained observer will tell by the smell of a sample its history, just as a tea-taster will pronounce an opinion on a sample of tea by its aroma.

147. To our readers who are engaged in the transport of the oil, the tests we are now about to describe are more useful. In the early part of this work, when speaking about the different kinds of petroleum, we in most cases gave their specific gravity, or, in other

words, a number which denotes the density of the liquid as compared with pure fresh water.

There are several instruments for obtaining the specific gravity of a liquid ; but for petroleum, that known as the Beaumé hydrometer is used, especially on the Continent.

This consists of a glass tube loaded at its lower end with mercury, and with a bulb blown in the middle. The stem is hollow, and has certain graduated divisions upon it. The principle of its construction is that which is one of the leading principles of naval architecture, and is that "a body floating in a liquid displaces a volume equal to itself in weight." In this hydrometer Beaumé took as a standard of density, a solution of 10 parts of salt in 90 of water. The instrument was immersed, and when at rest a mark was made at the point of flotation on the stem ; this is the zero. On immersing the instrument in pure water it floated more deeply immersed, and the point of flotation was marked ; the distance between these two points was divided into $10°$, and the division continued up to the top of the scale.

This graduation is entirely arbitrary and is not by any means a scientific method, but it has its advantages from a commercial point of view, and is extensively used in the petroleum trade. The marks on the stem of any hydrometer are only correct at a certain temperature, which is generally stated on the stem ; thus Beaumé's hydrometer is calculated for $15°$ C. or $59°$ F. ; for any other temperature the hydrometer will sink if the temperature is higher than this, and rise if lower by a very small amount, which for purposes on board ship may be neglected.

148. In a large tank filled with petroleum, the density will not be the same at the top and bottom. The heavier portion will occupy the lower part ; hence in obtaining the specific gravity of the oil, we must obtain samples from the bottom, the middle, and from the surface of the oil. The simplest way to do this is as follows :—

Procure an iron bar or heavy bolt about a foot long, or a hand-lead will do : to this lash firmly a good-sized bottle, holding about a quart, and have a small line made fast to the bar, which is merely for the purpose of sinking the bottle; have a good-fitting cork or wooden plug for the bottle, with a hole in the centre, or a staple on the top side, to which attach a length of small line (marline). The line on the iron rod should be divided roughly into feet or fathoms. Close the bottle and lower it into the filled tank, when it is on the bottom pull the small string, and the heavy oil will displace the air and fill the bottle ; it is then hauled up and the contents poured into the glass " immersion tube." By repeating the operation, only lowering the bottle to the middle of the tank, a sample of the oil at this depth is obtained.

149. To use the hydrometer we require a glass vessel—the immersion tube; this should be a glass tube about 14 inches long by 2 or 2½ inches diameter, and should hold at least enough to enable the hydro-

meter to be clear of the bottom. Empty the contents of the bottle into the immersion tube, and insert an ordinary glass thermometer, graduated to Centigrade, and the hydrometer. The latter must first be carefully wiped and free from any adhering dirt or grease. After a few minutes the hydrometer will float at rest. See that it does not touch the sides of the tube when taking the reading.

150. The readings of the hydrometer and the thermometer should be noted, and it is as well that a tank log book should be kept by one of the ship's officers, in which all particulars should be entered. It is, as a rule, only necessary to take the specific gravity when the tanks are just filled, and before signing bills of lading, as cases are known in which captains have signed a bill of lading specifying a certain grade of oil of a certain specific gravity, when, as a matter of fact, the cargo consisted of a totally different kind of oil. The master had not the means of making a few elementary tests, and, like many masters, signed just what the agent gave him.* Each day the temperature of the tanks should be noted—in the early morning and in the evening—and the height of the gauges (if any), or the distance of the oil from the top of the expansion chamber. Each morning and evening an officer should make an inspection of all the tanks, bilges, peaks, cofferdams, &c., and record his remarks in the tank log.

---

* The writer does not mean to impute carelessness to shipmasters. Too frequently the shipmaster is not permitted to use the discretion that the law supposes he ought to use when signing bills of lading.

151. Abel's tests.—Perhaps the most useful test is that by which the "flashing" and "firing" point of the oil is determined.

Liquids are divided into two classes, volatile liquids and fixed liquids.

In the former are comprised water, alcohol, ether, naphtha, petroleum, and others which have a tendency to pass into a state of vapour at ordinary temperatures. Among the fixed oils are those, such as olive, sperm, colza, &c., which remain unaltered at ordinary temperatures. Petroleum in its crude state contains, as we have seen, many exceedingly volatile compounds, consequently, in a large space like a storage tank or ship's tank, the formation of vapour is always going on ; or rather, so long as there is no change in temperature, a body of petroleum will give off vapour till the space surrounding it is saturated, or the vapour attains the state of maximum tension. An increase of temperature will lead to the production of more vapour. The case is somewhat analogous to an ordinary steam boiler. If the boiler is half filled with water and no heat applied, and free access of air is allowed, the water will very gradually pass off as vapour. If the water is heated by means of a fire underneath, the vapour is rapidly disengaged and passes into the air till all the water has disappeared. If the boiler, while containing water, is heated, and the opening to the external air closed, vapour is disengaged and the tension increased.

152. In a petroleum tank the disengagement of vapour occurs at even low temperature. On a hot

day, the vapour given off is greater, and if free communication exists with the external air all the petroleum would ultimately disappear. If the tank is
perfectly full, and there is no space for the accumulation of vapour, no disengagement of vapour takes
place, and, as petroleum vapour is exceedingly inflammable, a full tank is perfectly safe ; whereas a partially
full tank is dangerous, because the space not occupied
by the oil is filled with vapour, generally at its maximum tension.   Hence it is that the most dangerous
times in a petroleum vessel are those when loading
and discharging.   For this reason, too, ballast tanks
are a possible source of danger.   There is *always*
some small amount of oil in them, and this oil dis-
engages vapour. A foolish engineer using a naked
lamp to clear the "strum," will cause an explosion,
just as the engineer did on the *Ville de Calais*.
Petroleum vapour consists of very minute particles
of oil, so small that they are held in suspension in
space, just as, in fog, minute particles of water are
held in suspension in the air.   Now we know that a
mass of coal, or iron, or sulphur, cannot be consumed
readily in the open air, because there is not sufficient
oxygen to determine complete combustion ; but if we
have coal or iron or sulphur in a very minute state of
subdivision, a little friction, or any other method of
promoting a rise in temperature, enables the oxygen
to combine with these elements, with the rapid evolution of light and heat.   It is for this reason that fine
dry coaldust in mines becomes a distinct source of
danger.   A grain of wheat does not seem a very

likely subject to cause a fire, but grind that grain till it is in the form of a powder, and it readily catches fire ; for this reason special precautions are necessary in flour mills, cotton factories, &c.   Petroleum vapours, being merely particles of oil in a very minute state of subdivision, are in themselves harmless enough ; but when the oxygen of the atmosphere is likewise present, the presence of a flame enables the oxygen to combine with the vapour, and light, heat, and the production of gases, occur so rapidly that an explosion is the result.   An explosion is merely rapid combustion due to the circumstances of the case being favourable to the instantaneous combination of various compounds.

We might take one sample of petroleum, and expose it to the air, vapour would be given off, but at ordinary temperatures this vapour would be diluted so largely with air as to be harmless.   We might take another sample, and expose it to the air, and find that vapour was disengaged so fast that the approach of a flame would determine the combustion of the particles of petroleum vapour.   Hence it will be seen that the commercial value of a grade of oil depends in a very great measure upon at what temperature vapour is given off which, on the approach of a flame, will be ignited.   If we put some petroleum in a small vessel, and surround it with water, and apply heat, it will be noticed that at first on passing a lighted taper over the surface of the oil no ignition results.   As the oil becomes more heated, more vapour is given off, and at last a temperature is reached, when, on passing a flame over oil, the vapour suddenly ignites ; but there is only an instantaneous production of flame, or in other words the oil " flashes," and the temperature at which this occurs is the " flashing point." If we continue to apply heat, we shall, on passing the taper over the oil, obtain a succession of flashes, until at last the vapour is disengaged so rapidly that the flame is continuous, and lasts till all the oil is consumed. The temperature of the oil at which continued ignition takes place is called the " firing point." There are several methods of testing oil for its flashing and firing point, but the one officially adopted, both here and in America, and also used in commerical transactions, is known as Abel's tester (after Professor Sir F. Abel, the eminent chemist). We do not describe this instrument in detail, as it cannot be used on board ship (except in port). It is an instrument requiring very great care and skill in its manipulation, and it also involves the use of a pendulum. We, however, give a sketch of a rough modification of the Abel tester which will give fairly accurate results, sufficient at any rate for use on board ship. The writer has used such an instrument made by himself, and found the indications come out within a degree or so of the official test. It must be remembered that the writer's object is not to give the method of accurate determination, which can only be arrived at by an expert, but merely a practical method by which the captain can satisfy

himself that the oil he signs for in the bill of lading is
what he has on board.   This is all the more necessary
as the custom of making captains "sign for quality"

FIG. 65.

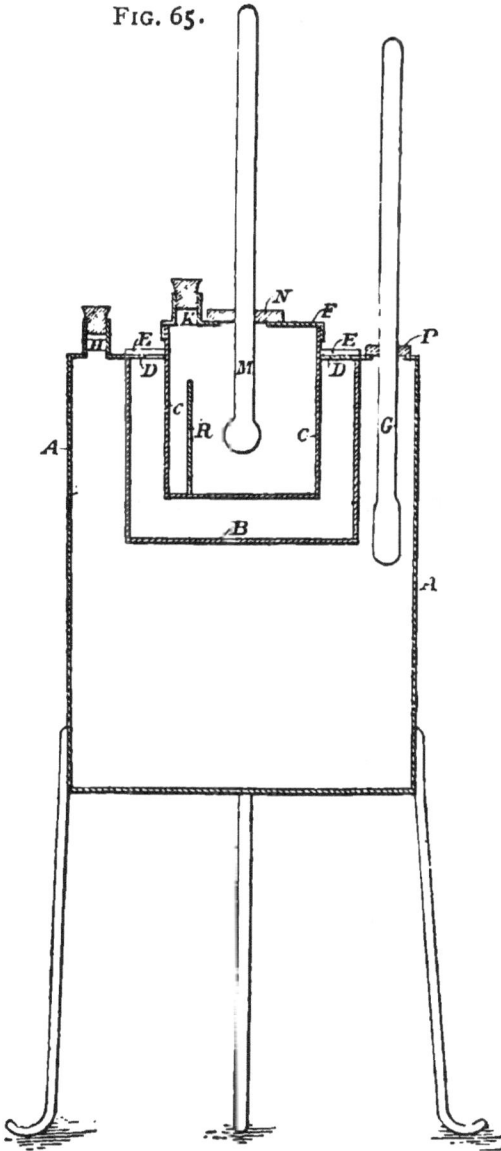

(in connection with other cargoes) is becoming very common in many parts of the world.

153. The apparatus is made of sheet brass about No. 22 B.W.G. thick. There are three cylinders, A, B, and C. A is $5\frac{1}{2}$ inches diameter by $5\frac{3}{4}$ inches high; B is 3 inches diameter by $2\frac{1}{2}$ high; C is 2 inches diameter by $2\frac{2}{10}$ high. A plate D is fitted over the two cylinders A and B, and a space is cut out sufficient to allow the cylinder C to be inserted; the cylinder C is fitted with a flange E, $\frac{1}{2}$-inch wide, and $\frac{3}{8}$-inch from the top; a cover F is made to fit tightly the cylinder C. Two holes are made in the bath, one for the insertion of a thermometer G, and the other H for filling and emptying the vessel.

In the cover of C is a small spout K, $\frac{1}{2}$-inch high by $\frac{1}{4}$-inch diameter, this is closed by means of a small wooden plug. A hole $\frac{1}{2}$-inch diameter is also made in the cover for the insertion of a thermometer, M. The outer cylinder A is mounted on three legs, a small spirit lamp is provided, and one or two tapers.

The thermometer M has a tight-fitting ring of vulcanised indiarubber round it, N, which holds it in position, and at such a distance that the middle of the bulb is $1\frac{1}{2}$ inch from the top cover. The thermometer G has a similar ring of vulcanised indiarubber, P.

To the bottom of the bath C is soldered a piece of wire R, which is $1\frac{1}{2}$ inches from the bottom; this acts as a gauge, and denotes the height that the oil is to fill the bath.

To use it to test the quality of an oil, proceed as

P

follows :—Place the apparatus where it will not be subject to draughts. Fill the bath about four-fifths with warm water, and insert the thermometers, which should read up to 190° F., and light the spirit lamp and place it underneath. Apply heat till the temperature of the bath is 130°, then remove the lamp and introduce the inner cylinder containing the sample of the oil, which should just touch the top of the gauge R. The thermometer being fixed in the inner cylinder (a piece of vulcanised indiarubber, N, fitting tightly round the stem, is a capital support for it ; the bulb should be 1½ inch from the top of the lid), the temperature of the oil will increase. When the thermometer shows 66°, the wooden stopper in the tube is withdrawn with the left hand, and a lighted taper held in the right hand is just passed over the mouth of the tube, this is done with each rise of one degree, till at length, on passing the lighted taper over the tube, a flash is seen, the temperature is noted, and this denotes the flashing point of the oil. By further heating the oil we should reach a temperature when the taper passing over the tube would cause a permanent flame to issue, this would be the "firing point."

By the Petroleum Act of 1879, the flashing point of petroleum must not be less than 73° F.

The above method of testing is really a modification of Abel's, but without the refinements. In using such a simple piece of apparatus little skill is required, and the apparatus itself can be made easily on board.

154. As regards the foregoing tests, the sp. gr.

test gives us an indication of the denser compounds present, but says nothing as to the inflammability of the oil; this latter is, as we have seen, sufficiently indicated for practical purposes by the flash and firing test.

On page 20 we referred to the fact that the specific gravity of an oil only held good for that particular temperature. In finding the quantity of oil contained in a tank of known capacity, the temperature and the specific gravity of the oil have a most important influence.

All calculations relating to petroleum are generally made on the basis of 15° C., equal to 59° F., which is the normal temperature of North America, England, and South Russia.

155. On page 197 we gave a table showing the coefficient of expansion for oil of various specific gravity at the normal temperature of 15° C. We shall now show how to calculate the weight of oil in a tank.

To find the quantity of oil in a tank, knowing the number of cubic feet in it :—Find the specific gravity of the oil, and multiply it by the weight of 1 cubic foot of pure water, equal to 62·415 pounds; multiply this by the number of cubic feet in the tank, and the product will be the weight of the oil in the tank in pounds; this, divided by 2240, will give tons.

156. Having given the specific gravity of an oil at a certain temperature, to reduce it to the specific gravity it would have at the normal temperature of 15° C. :—Take the temperature of the oil in degrees C.

If the temperature is higher than 15°, multiply the excess by the coefficient of expansion ; add this product to the specific gravity, and the sum will be the specific gravity of the oil at the normal temperature. If the temperature of the oil is less than 15°, multiply the defect by the coefficient of expansion and subtract the result from the specific gravity. The remainder will be the specific gravity of the oil at the normal temperature.

CHAPTER VI.

THE LIGHTING OF PETROLEUM STEAMERS BY ELECTRICITY.

157. PETROLEUM steamers should always be lit by electricity, and no other form of illumination should on any account be permitted. At the same time, even with the electric light considerable care is required, as under quite possible conditions an explosion in a tank could easily be arranged with the aid of an electric current. The nautical mind has firmly "caught on" to the idea that the electric light is at all times absolutely safe.

We need only mention that in a very favourite form of "oil tester" (Siebolt's), the electric spark is used to determine the flashing point, instead of the small flame we have described. Submarine mines are exploded by the heating of a small wire caused by the passage of an electric current.

It will readily be conceived that under certain circumstances an electric light installation may become a source of danger, especially if the ship is wired on the single-wire system, with the hull of the ship as the return lead ; therefore the writer is of opinion that petroleum steamers should in all cases be wired on the 2-wire system. This of course involves twice the amount of cable and lead, and is more expensive, but it is very much safer. It may be rash to condemn the single-wire system, when we see it used extensively in the largest class of mail steamers, but in the case of petroleum steamers there cannot be two opinions on the subject.

158. The writer consulted a well-known electrical engineer on this point, and this is what he says :— " When lighting ships in the old days, a great many of the leading companies adopted the single-wire system, and even the Admiralty tried it, on board H.M.S. *Polyphemus.* This last installation is said to have succeeded well, but it is doubtful, as, generally, what is called a success, is only partial ; that is to say, if the installation breaks down " *not too often,*" it is called a success, and the few breaks down are supposed to be inseparable from any electric light scheme ; this is quite an erroneous impression. The single-wire system consists in the using of an insulated conductor as an outgoing lead, and the ship's skin as a return. The great disadvantages are, that salts form on the joint between the lamp wires and the ship's skin, and introduce an electrical resistance, causing the lamps to burn dimly and unevenly. In

testing such currents, all the lamps require to be individually disconnected, as, of course, each lamp is to earth. This is a great source of trouble ; and further, if one earth occurs in such a circuit, a certain number, if not all, of the lamps will be short-circuited, and go out."

159. There are, as might be expected, many vessels fitted with electric light installations which will not work, or give endless trouble if they do. Such plants have usually been laid down by marine superintendents and others, whose knowledge of electro-technics is generally nil. Such plants are extremely valuable, however, as they teach one what things to avoid. Sometimes a vessel is " lit throughout by the electric light," and the dynamo engine takes steam from the winch steam pipes. If the winches are running, the lights bob up and down, especially if, as is generally the case, the governor is not used. Very often, too, the steam-pipe is led along the upper deck and exposed to the weather, and if the vessel happens to ship a few sprays, the steam is condensed and the light goes out for a time. Sometimes, too, the dynamo is placed immediately under a deck ventilator, and water and dirt drop on the armature. Very often, too, main leads are led and attached to the main boiler casing, where they are subject to a heat of 250°. These are all valuable examples of "how not to do it." It is not uncommon to see main leads taken through bunkers and passed through holes in the iron frames, and jammed up with a wooden wedge. In time coal

and the trimmer's shovel rub off the insulation or destroy the wire sheathing, and the electric light is anathematised. On the other hand, in vessels where the plant has been put down under competent superintendence, the electric light gives the happiest results and for ship use it is, without doubt, *the* light. It is often said that an electric-light plant can be put down that will require no skilled attention. In many vessels the donkeyman runs the show with more or less (generally less) satisfactory results. The truth is that to obtain success with a plant, a certain amount of skilled attention and electrical knowledge is absolutely necessary. The writer is not aware by what process of mental evolution a ship's donkeyman becomes possessed of the requisite technical knowledge. We have known more than one armature " burnt out," and more than one voltmeter ruined, through the excessive looking-after, or want of it, indulged in by the donkeyman. It is true that little knowledge is required to start an engine and dynamo and keep it going, but it is when a short circuit occurs, and when faults develop themselves, that something more than a donkeyman's knowledge is required.

160. Engines.—Tangye, Lindley and Burnett, Allen & Co., and Willans & Co., make splendid engines for direct driving.

Many designs of high-speed engines are made by Tangye's, Limited, Birmingham, for electric lighting purposes, which include engines of various horizontal and vertical types, "column," compound, triple expansion, and many others too numerous to mention, but which may be seen on most of the leading steam fleets throughout the world, including the Peninsular and Oriental, North German Lloyd, Cunard, White Star, and many others.

The type selected, see Fig. 66, is that known as the "Archer" coupled compound, and, in connection with the dynamos of the Anglo-American Brush Corporation, has been supplied to all the last new boats built for the P. and O. Company. The engines are intended for a speed of 200 revolutions per minute, working under a steam pressure of 150 lbs. The high-pressure cylinder is 18 inches diameter, low-pressure 16 inches, both with a stroke of 10 inches. The piston rods and crosshead are of steel, forged together, steel valve spindles and crankshaft with throws slotted out and machined all over. The connecting rods are of wrought iron fitted with marine heads for crank pin-ends. The valve spindles are guided and the joints fitted with wedge adjustments. The engine is controlled by a Tangye patent governor and fitted with gun-metal stop-valve. All the bearings are of extra large size, and the lubrication is so arranged that all parts can be oiled whilst working. The cylinders are lagged with sheet steel and fitted with relief valves, asbestos packed drain cocks, and gun-metal sight-feed lubricator. The exhaust from the low pressure cylinders is conveyed to the ship's condenser. It will be observed that the plant is very compact. This type of engine and dynamo gives very great satisfaction and can be strongly recommended.

Electrical machinery, from its very nature, demands the best and most skilful fitting. When getting out the plans of a vessel, a good-sized place should be set apart for the dynamo and its driving engine.

161. Do not act on the idea that a dynamo can be " shoved anywhere "—this is often done, with the result that the whole installation becomes a constant source of annoyance. There are so many good makers of dynamos that it might be invidious to mention any particular maker, but some firms make a speciality of ship lighting. *The* dynamo is, without doubt, the " Edison-Hopkinson " and the " Manchester " by Mather and Platt. Holmes of Newcastle, the Anglo-American Brush Company, Woodhouse and Rawson, the India Rubber and Gutta Percha Telegraph Company, of Silvertown, Crompton and Company, all manufacture high-class dynamos which will be found to give the fullest satisfaction. For ship lighting low voltage will of course be used, either 55 or 60, or not above 110 volts. Higher volts than these are not recommended. The writer is not inclined to go above 50–60 volts, as, in lamps intended for higher voltage, the carbon filaments are so thin that the continued vibrations, even with tri-compound engines, soon cause them to break, long before the life of the lamp, from an electrical point of view, is exhausted. The filament of a low voltage lamp, being much

thicker, is more substantial. In estimating the power to be developed by the main boilers, due provision should always be made for driving the dynamo engine. This is mentioned, as it is not always done ; many engineers and owners thinking that the addition of another engine, to be supplied with steam from an already overburdened boiler, makes no difference. It does ; and if the exhaust from the dynamo engine goes up the chimney instead of into the main condenser, extra feed has to be put on.

162. The dynamo-room should be as far as possible from the main engines and boilers ; and, what is of still more importance, it should not be within 50 feet of the position of the standard compass and chronometer. The deck of the dynamo-room should be of iron.

Do not put a dynamo in the engine-room, with the mistaken idea that the engineer of the watch can look after it. With the modern tri-compound engines an engineer has all his watch fully employed without giving him another couple of machines to keep in order. Besides, the engine-room is a bad place. The temperature is always very high, which is bad for the insulation. If a gland gives out at sea the engine-room is likely to become full of damp hot vapour, which will in time ruin a good dynamo. When coaling, coaldust always gets into the engine-room, and settles on everything ; and coaldust and moisture gradually collect on the armature, and it is merely a question of time when successive accumulations of dirt seriously impair the efficiency. In vessels of the latest design a platform is built in the after-end of the engine-room, in which the dynamo and engine are placed ; and if a small electromotor is supplied to drive a ventilating fan, the place can be kept nice and cool.

Do not place the dynamo immediately under a deck ventilator; nor the engine either. The steam pipe for the engine should lead direct from the main boiler, with a branch from the donkey boiler and should *not* be taken on deck, but should be led along under shelter to the dynamo room. In heavy weather trouble is frequently caused through sprays and small seas coming on deck, condensing the steam in the steam pipe, and causing the lamps to go out, or else become so dim as to be useless. A governor should be supplied, and if supplied should be used. The Acme and Tangye's and Pickering governors give excellent results. At sea governors require a little attention to keep clean, and if they are not kept in good working order, and free from accumulations of greasy deposit, they are apt to become useless.

If a dynamo of the horseshoe type is selected, that one should be preferred in which the pole-pieces are uppermost. Bearings for the armature spindle should be very long, and bushed with phosphor bronze ; spare bushes should be included in the order.

Stipulate that the commutator sections should be insulated with mica.

This is very important. There are, or were, a lot of dynamos—"cheap," of course—built for ships' use with the commutator sections insulated with vulcanite. Generally speaking, no one on board ship knew anything about it, and hence these machines were palmed off on the owner. Having decided upon the dynamo see that a proper place is prepared for it ; a separate room, if possible.

If the dynamo and engine are coupled for driving direct, both should be mounted upon a cast-iron bed-plate, which should be bolted down to two wooden beams of ample cross section. The whole is secured to the deck by means of through bolts. If the dynamo is driven by belting, either cotton or leather, rails should be similarly fixed, and screw-tightening gear supplied to regulate the tension on the belt.

163. If the engine and dynamo are coupled direct by means of a flexible coupling, they should perhaps be placed athwartships, as the rolling of the vessel allows the shaft to have a little longitudinal motion in its bearings, which is not a disadvantage. If belting is used, the dynamo and engine should be placed fore and aft.

Friction driving, if of a good type, can be used very advantageously for running dynamos. Raworth's has been used with very great success. If belting is used, cotton and leather are usually employed. The former makes very smooth running, but is apt to chafe in time, and the splices give trouble unless made by a skilful sailor.

In running dynamos with leather belting, see that the "grain" side of the leather is next to the pulley.

It is not at all a bad dodge to cover a pulley with leather neatly and carefully sown on. A belt will do more work and wear longer on a leather-covered pulley than on any other. Short belts should be tighter than long ones.

In order to have belts run well, they should be perfectly straight, and have but one laced joint. Belts and pulleys should be kept clean and free from accumulations of dust and grease, and particularly of lubricating oils, some of which permanently injure the leather. On no account use castor oil; it rots the leather. The most effective tightener is the weight of the belt on its slack side, which increases adhesion by increasing circumferential contact with the pulleys.

In using ordinary leather belting, see that the joint is well made, if it is not, every time the joint goes over the dynamo pulley a sudden jump takes place in the current, causing the light to flicker.

164. In running dynamos it must not be forgotten that the practice which obtains in slow-speed machinery is not applicable. In a dynamo we have a machine with only one moving part, the armature, and this revolves at speeds in different machines from 200 up to 1500 revolution per minute. In ordinary dynamos of good design and make, due provision is made for good lubrication, which should be automatic and continuous. Mineral oil makes the best lubricant;

if too thick, it may be thinned by the addition of a little paraffin. Suppose we are about to light up for the evening: see that the oil cups are full and working properly, and give a drop of oil every 30 seconds. Carefully examine the armature, and see that no dirt, pieces of waste, &c., are lodged anywhere. See that the brushes are properly fixed in the holders, and that they press evenly on the commutator.

The engine having been drained and warmed up, can be turned over to clear it of any condensed water. If the dynamo and engine are coupled together, do not let the brushes rest on the commutator till ready to light up; see that the nuts and brasses are not slack at any part. Start the engine at moderate speed, and turn the brushes on the commutator. We are now running the dynamo on "open circuit," and are producing a current which is exciting the field-magnets—the presence of this current should be manifested by the "Pilot lamp." When the current is being generated fast enough the circuits can be switched on one by one, and the stop valve suitably adjusted till the dynamo is at its full load. If a good governor is fitted it will not be necessary to touch the stop valve any further till shutting-down time. As the lights in the cabins are turned out, this of course eases the load, and the engine runs faster, and unless the governor is good and in good working order, an excessive brightness in the other lamps results, which diminishes their life. Do not run lamps at a brightness more than can be comfortably borne by the eye.

Under normal conditions there should be no sparking
at the brushes, if there is it may generally be cor-
rected or minimised by turning the brush-holders in
the "rocker" backwards and forwards, till a position
is found when the sparking is least. It may arise
from the brushes themselves not bedding firmly, but
jumping; the presence of dirt, coaldust, grease, &c.
will all produce sparking. One result of sparking is
that "flats" are produced in the commutator bars, and
unless it is remedied it becomes worse and worse. A
good practice is to lubricate the commutator with a
very small amount of mineral oil. Great attention is
necessary to keep the brushes and commutator in
good working order. After every run, carefully wipe
off all copper-dust, grease, &c., and give the commutator a rub with the finest emery paper.

165. While the dynamo is running feel the field-
magnets occasionally; they will become warm, but
should never be hot, and all parts should be of the
same temperature; if one magnet coil feels warmer
than another, this shows that there is something
wrong with the cold coil—probably it will be the
presence of a "short circuit." Stop the machine and
carefully examine the insulation, and see that the
small wires have not in any part had the insulation
rubbed off, or that they are not touching the framing,
or any other wire. It may happen, too, that the
current while passing round one field-magnet, owing
to defective insulation, passes also into the magnet
itself. To see if this is the case, run the dynamo at a

good speed, and take a short piece of insulated wire and join it on to one terminal, and with the other end touch the shaft or some part of the bare iron of the yoke ; on removing the end of the wire, if there is a spark, it shows the insulation on the field coils is defective, repeat the test with the other terminal.   If no spark is obtained the insulation is perfect.   It may happen, too, that on running the dynamo no current is obtained, this may be caused by some of the connections on the machine being broken.   After running the dynamo carefully, feel each coil of the armature ; they should all be of the same temperature, if one part is hotter than another, it shows a "fault"—one coil is doing more work than the other.   Carefully examine the commutator section and see that each section is perfectly insulated from its neighbour.

In general little can be done in the way of effecting repairs in the dynamo on board ship.   After long running all commutators gradually assume an oval shape, and sparking is produced.   The only thing is to have it turned up in a lathe by an experienced man.

A canvas cover should be made to cover the engine and dynamo, and after every run the latter should be carefully wiped down, and brushes, bearings, &c., examined ; governors require careful cleaning and adjusting, and finally the covers should be put on and stropped or laced at the lower edges to a jackstay on the bed plate.   If this care is adopted much trouble and annoyance will be saved.   To leave a

Q

fast-speed engine and dynamo unprotected, as is too often done while the ship is coaling, is simply most reprehensible, and shows either carelessness or ignorance on the part of the engineer in charge. If the main switchboard is open a canvas cover should also be provided for it.

166. Probably the most important part of an installation is the wiring, and unless this is done well and carefully, faults will occur which may cause positive danger. The separate departments of the ship should each be served with a separate circuit, and in petroleum steamers *the hull of the ship should in no case be used as the return load.* In specifying the size of the conductor see that ample section area is provided and do not let false notions of economy prevent the employment of fair-sized conductors. For example, do not use single wire of No. 22 B.W.G. for running a single lamp on. Although such a wire will safely carry the amount of current necessary, much better use No. 18.

The conductors of the main circuit should be of sufficient sectional area to carry at least half as much again of current, as they would have to carry under normal conditions. It may happen, for example, that the circuit for the forecastle, passing through the main bunkers, might be damaged and rendered useless for some time. A competent electrical engineer, perhaps with some spare lead on board, would run the lamps off another circuit; of course he would first satisfy himself that it could be done safely.

Ships find it extremely difficult to obtain electrical supplies abroad, and hence the conductors should be of ample size, and provision should be made, as far as practicable, to run one set of lamps off another circuit.   Of course it is understood that the copper has at least 96 per cent. conductivity.

167. The insulation of conductors for ships' use should be of the very highest class, and the contractor should specify the maker.   All the cables and leads should be insulated with pure indiarubber, then vulcanised indiarubber, then indiarubber tape, then a stout covering of braided hemp, and finally a coat of preservative compound.   The conductor should be tinned ; and when ordering lengths of cables and leads, specify that they shall have an insulation of not less than 300 meg-ohms per statute mile at 60° F.

Cables and leads should not be hidden away behind the panelling of cabins, but should be enclosed in casing which is easily accessible.   Two or more conductors should not be laid in the same groove of the casing ; do not attach leads to panelling by metal staples.

The flexible leads for hand lamps or the group lamps should be additionally protected for use in oil steamships, as petroleum tends to soften and destroy the insulation.   In using hand lamps in an oil tank very great care is requisite to prevent sparking.

If the ship is wired on the single-wire system the iron beams, and hinges, frames, &c., all form part of the "return lead " to the dynamo.   If, in using a hand lamp in a tank, the insulation of the leads is not perfect, the current passes into the brass framing of the lamp, and if the latter be allowed to touch the naked iron of the tank, the current, instead of going through the carbon filament, will take the easier path through the iron of the ship; on removing the lamp the self-induction on breaking the circuit manifests itself as a spark. Should the mixture of petroleum vapour and air in the tank happen to be of the right proportion, an explosion would result.

The writer recommends that the casing of the hand-lamps should be covered over with dry spun yarn soaked in paraffin wax. All the brasswork of the lamp which is likely to touch anywhere should be thus protected.

It is also a good practice to "serve" the flexible leads with clean white spun yarn, as the ordinary protective covering soon becomes worn by friction over sharp tank edges.

168. The glow lamps, as they are called, are generally of 8 and 16 candle-power. They are usually "capped," and fixed in a socket. The writer prefers looped lamps, as in the former the lamp is apt to work loose in its plaster of Paris setting, and become useless. There are many different ways of fixing lamps, but the chief thing is to guard against vibration. In the capped lamps, the socket grasps the cap firmly, and any shock is transmitted to the lamp; whereas, in the looped lamps, there is a spring which takes up any vibration and lessens it very

considerably. Woodhouse and Rawson make some excellent lamp holders, very suitable for ships' use, as are also their well-known slate-based G switches. The lamps are of course run "in parallel." Arc lamps are not suitable for ordinary vessels unless a skilled electrician is carried, and for oil steamships they are simply out of the question.

Roughly speaking, allow 7–8 60-watt lamps, requiring one horse-power. In choosing a dynamo, the one which gives its full output at a low speed, say not exceeding 280 revolutions per minute, is to be preferred. At this speed it can be driven direct by a suitable engine.

For illuminating the deck at night time, when loading or discharging, the "Sunbeam" lamp is very useful. It should be placed in a silvered copper reflector, and of the most substantial make. These lamps can be obtained of any desired candle-power, but the 200–300 candle-power lamp will be found very suitable. At this candle-power the lamp takes a current of 10 to 15·5 ampères at an electromotive-force of 50 volts.

Bulkhead lamps should be fitted in the fore-castle, galley, and engine rooms, in place of the usual lamp, the former are more substantial, and can be secured against any interference.

Switches and cut-outs should be of a good substantial make, and should be placed in situations where no moisture can get at them.

The main switch-board should be of slate, and

enclosed in a cupboard, so as to exclude coal-dust and dirt. Of course when running it is kept open. The switches and cut-outs should be of solid, simple and substantial construction, and of such a character as will admit of their being easily kept clean. A volt meter and an ammeter are sometimes supplied, but this is not necessary unless a skilled electrician is carried ; besides, the indications they give are seldom correct, and they require frequent calibrating.

If a competent man is carried it would be well to supply a few Leclanché cells, a galvanometer and a box of resistance coils, and a " bridge." Tests for insulation and conductivity should be made at frequent intervals ; if this is done, faults will be detected in time, and much time and worry saved.

Owners should engage an electrician to test the circuits each voyage, and furnish a report ; " a stitch in time " is a proverb very applicable to petroleum steamers.

Nothing is said about secondary batteries used in conjunction with a dynamo for lighting a ship. They require very skilled attention, and are in fact unsuitable for ships' use.

With the current of an electromotive force of 50 to 100 volts, such as are generally used on board ship, little danger need be apprehended if the naked terminals are touched by hand, although an unpleasant shock may be given.

169. It is very important that the disturbance of the compass due to the electric light leads should be

eliminated, and owners are advised to insert a clause in the building contract that no such disturbance shall take place. Captains will do well to take advantage of their ship being moored in dock or lying at a steady anchor, to note the deflection (if any) produced by the dynamo when running under different conditions.

Supposing at sea it is necessary to take a pair of leads near the compass, using a continuous current machine, the wires in the vicinity of the compass should be twisted together ; even then great watch-fulness is requisite on the part of the captain. The writer was recently on board a large mail steamer ; on the upper bridge, situated a few feet away from the compass, was a projector. After an inspection of the electric light plant (the ship was wired on the one-wire system, and used a continuous current machine), the writer made a bet that when the arc light was being used a deflection was produced on the compass. The bet was taken up, and won by the writer. This " yarn " is told, not to show any smartness on the part of the writer, but merely to indicate the kind of professional advice that is sometimes given, even to the best-known and richest companies. Here was a ship "with all the latest improvements," built under an unusually close survey, both of Lloyd's and the Government, and yet it was considered quite the right thing to place a projector in the immediate vicinity of the compass.

If the dynamo used is a continuous current machine, care must be taken that the main leads do not come anywhere near the compass. In a valuable paper, lately read by Sir W. Thomson, F.R.S., before the Society of Engineers, he showed that a cable carrying 100 ampères at a distance of 33 feet from the compass, produced a deflection of 7·6 degrees of the needle. If an alternate current machine is used, this precaution is not necessary; but as a rule alternate current machines require the use of high volts, and as a matter of fact are seldom if ever used on board ship.

CHAPTER VII.

USEFUL DATA, TABLES, ETC.

170-1. Pumping Engines—172. Weight of Cast-iron Pipes—173. Weight of Oil contained in a Length of Piping—174. Pressure of a Head of Water—175. Air Pressures—176. Stowage of Coals—Stowage of Oil—Stowage of Tallow—Stowage of Waste—177. Capacity of a Tank—178. Casks—179. Tank Car—180. Salt-water Data — 181. Fresh-water Data—182. Useful Factors—183. Specific Gravity of Seas—184. Beaumé's Natural Specific Gravity—185. Thermometer Scales—186. Vocabulary.

170. Pumping engines :—

G = Number of gallons discharged per minute.

C = Number of cubic feet discharged per minute.

D = diameter of pump in inches.

L = length of stroke in feet.

N = number of strokes per minute.

H = horse-power to raise G gallons or C cubic feet per minute.

h = height beyond is to be lifted.

G = ·03401 N L D².

C = ·005456 N L D².

$$D = \sqrt{\frac{29 \cdot 4\,G}{N\,L}}, \text{ or } \sqrt{\frac{183 \cdot 3\,C}{N\,L}}.$$

$$H = \frac{N\,L\,D^2\,h}{97020} \text{ or } \frac{c\,h}{15557}.$$

—(Mackrow.)

171. To find the diameter of a single-acting pump :—

L = length of stroke in feet.

G = number of gallons to be delivered per minute.

F = number of cubic feet to be delivered per minute.

N = number of strokes per minute.

D = diameter of pump in inches.

$F = \cdot 00545 \; D^2 \, L \, N.$

$G = \cdot 034 \; D^2 \, L \, N.$

$$D = \sqrt{\frac{G}{\cdot 034 \, L \, N}}.$$

$$D = \sqrt{\frac{F}{\cdot 00545 \, L \, N}}.$$

N.B.—These formulæ give the nett diameter of the pump plunger; it is usual to increase the area of the plunger one-fourth to allow for leakage. — MOLES-WORTH.

To find the cubic capacity of a cylinder—

d = diameter.

l = length.

$d^2 \times \cdot 7854$ = area of circle.

$d^2 \times \cdot 7884 \times l$ = cubic contents.

If d and l are in inches it is better to express them as decimals of a foot, and the cubic contents are then in feet.

Example.—A ballast tank is 90 feet long, 30 feet in breadth, and 3 feet deep. How long will it take to pump it out by the donkey-pump?

Diameter of piston 9 inches, stroke 14 inches.

Revolutions 120 per minute double acting, and ¾ full each stroke.

$90 \times 30 \times 3 = 8110$ cubic feet in tank.

$= 231 \cdot 5$ tons nearly.

$$\frac{9^2 \times \cdot 7854 \times 14 \times 120 \times 3}{1728 \times 4} = 92 \cdot 775 \text{ cubic feet.}$$

$$\frac{8100}{92 \cdot 7} = 87 \cdot 3 \text{ minutes.}$$

Having given the diameter (d) of the pump, the stroke of the pump (l), the number of strokes (n) per minute, and the fraction (k) that the pump is filled each stroke, to find the quantity pumped in a given time—

$$a^2 \times \cdot 7854 \times l \times k \times n \times 60.$$

A pump is 6 inches diameter and 15 inches stroke, and makes 70 strokes per minute; the pump being $\frac{4}{5}$ full each stroke.   What is the amount pumped per hour?

$$\frac{6''^{2} \times \cdot 7854 \times 15'' \times 4/5 \times 70 \times 60}{1728 \times 35} = 23 \cdot 56 \text{ tons.}$$

We divide by 1728, because 1728 cubic inches = 1 cubic foot and by 35, because 35 cubic feet of sea-water = 1 ton.

The process may be shortened by the introduction of factors, thus—

$$\frac{6 \times 6 \times \cdot 7854 \times \overset{5}{15} \times 4 \times \overset{2}{70} \times \overset{3}{60}}{\underset{2}{12} \times \underset{2}{12} \times \underset{4}{12} \times \underset{1}{5} \times \underset{1}{35}}$$

thus we have

$$3 \times 2 \times 5 \times \cdot 7854 = 23 \cdot 56 \text{ tons.}$$

Useful numbers for pumps :—

 D = diameter of pump in inches.
 S = stroke of pump in inches.
 $D^2 S \times \cdot 7854$ = cubic inches.
 $D^2 S \times \cdot 002833$ = gallons.
 $D^2 S \times \cdot 0004545$ = cubic feet.
 $D^2 S \times \cdot 02833$ = lbs. fresh water.

—(MOLESWORTH.)

172. To find the weight of cast-iron pipes :—

> D = diameter outside in inches.
> d = diameter inside, or bore in inches.
> W = weight of 1 yard of pipe in lbs.
> W = $7 \cdot 35$ (D² − d^2).
> The weight of two flanges = about 1 foot of pipe.

—(MOLESWORTH.)

A 6-inch iron pipe 9 feet long is composed of metal $\frac{7}{16}$ inches thick and weighs, 2 cwt. 2 qr. It will stand a head of water 300 feet high, and a pressure of 130 lbs per square inch.

173. To find the weight of oil contained in a length of piping (rough rule). Square the diameter in inches, and the result will be the weight in pounds avoirdupois in a 3 foot length.

174. Table of the pressure of water at different heads. H = head in feet, P = pressure in lbs. per square foot, p = pressure in lbs. per square inch.

H.	P.	$p.$	H.	P.	$p.$	H.	P.	$p.$
1	62·4	·4333	5	312·0	2·1666	30	1872·	13·000
1·25	78·0	·5416	6	374·4	2·6000	40	2496·	17·3333
1·5	93·6	·6500	7	436·8	3·0333	50	3120·	21·6666
1.75	109·2	·7583	8	499·2	3·4666	60	3744·	26·0000
2	124·8	·8666	9	561·6	3·9000	70	4368·	30·3333
3	187·2	1·3000	10	624·0	4·3333	80	3992·	34·6666
4	249·6	1·7333	20	1248·0	8·6666	90	5616·	39·0000

—(MACKROW).

175. Table of the force of air.

Velocity of the Air.		Force per square foot in pounds.	Velocity of the Air.		Force per square foot in pounds.
Miles per hour.	Feet passing per second through one foot area.		Miles per hour.	Feet passing per second through one foot area.	
I	1·47	·005	30	44·01	4·429
2	2·93	·020	35	51·34	6·027
3	4·40	·044	40	58·68	7·873
4	5·87	·079	45	66·01	9·963
5	7·33	·123	50	73·35	12·300
10	14·67	·492	60	88·02	17·715
15	22·00	1·107	80	117·36	31·490
20	29·34	1·968	100	146·70	49·200
25	36·67	3·075			

Atmospheric pressure, &c.—Atmospheric pressure = 14·7 lbs. per square inch, or 2116·4 lbs. per square foot = the pressure by weight of a column of water 34 feet high = a column of mercury 30 inches high.

Expansion of air for every degree of increase of temperature (F.) = $\frac{1}{459}$th of its volume.

T = higher temperature.
t = lower temperature.
V = volume at the higher temperature.
v = volume at the lower temperature.

$$V = v \times \frac{459 \times T}{459 \times t}$$

100 cubic feet of air at 32° = 136·7 cubic feet at 212°. Density of air to that of water as 1·815.

176. Coals :—

Cubic feet of coal, multiplied by ·0345, gives tons

  ,,   ,,   ·023, ,, tons of stowage.

Oil :—

One cubic foot = 6·23 gallons.

,,   ,,   = 58 lbs.

lbs. divided by 9 = gallons approximate.

Tallow :—

One cubic foot = 59 lbs.

Waste :—

One cubic foot = 11 lbs.

177. To find the capacity of a tank :—

l = length in feet.
b = breadth in feet.
d = depth in feet.
$l \times b \times d$ = No. of cubic feet in tank.
$l \times b \times d \times 62\cdot4$ = No. of lb. of fresh water.
$$\frac{l \times b \times d \times 62\cdot4}{2240} = \text{No. of tons of fresh water.}$$

For sea water use 64·05 instead of 62·4.

178. Capacity of casks :—

D, d = inside diameter at the heads.
M = inside diameter at the bung.
L = length (all in inches).

Capacity in Imperial gallons :—

$$= \cdot0014162 \,(D\,d\,M^2).$$

The buoyancy in lbs. equals 10 times the capacity in gallons *minus* the weight of the cask itself.

Petroleum in barrels.—An ordinary barrel is 33 inches long and 25 inches in diameter at the middle, and weighs when full about 400 lbs. Such a cask holds about 42 imperial gallons, and its own weight is 64 lbs., or about one-fifth of that of the oil it contains. It is estimated that in stowing this cargo, three and a half casks take up on the average a ton of 50 cubic feet, and it would therefore, occupy 80 cubic feet per ton deadweight. But as most modern three-deck steamers, fitted with water ballast in the holds in the usual way, cannot be brought to their load draught with cargoes occupying more than about 50 cubic feet to the ton, it will be seen that a great loss of cargo-carrying power is sustained by shipping petroleum in such vessels in casks, through the fact that the whole of the available cargo space is occupied long before the vessel is brought down to her proper load draught.—(MARTELL.)

Allow 75 lbs. as the weight of cask, of which 12 lbs. is the weight of the hoops.

In stowing petroleum in casks, $3\frac{1}{2}$ casks = 1 ton of 50 cubic feet, therefore one ton of petroleum in casks occupies 80 cubic feet.

According to the New York Produce Exchange barrels are to be made of well seasoned white oak timber, and shall be hoopel not lighter than as follows : Either with six iron hoops, the head hoop $1\frac{3}{4}$ inches wide, No. 16 gauge (English Standard) the quarter hoop $1\frac{1}{2}$ inches wide, No. 17 gauge, and the bilge hoop $1\frac{3}{4}$ inches wide, No. 16 gauge ; or, with eight iron hoops, the head hoop, $1\frac{3}{4}$ inches wide No. 17 gauge, the collar hoop $1\frac{1}{4}$ inches wide, No. 17 gauge, the quarter hoop $1\frac{1}{2}$ inches, wide, No. 18 gauge, and the bilge hoop $1\frac{1}{2}$ inches wide, No. 18 gauge. But all old barrels of which the gross weight is less than 395 pounds may be hooped with six hoops $1\frac{1}{2}$ inches wide, excepting the chime hoop, which shall be $1\frac{3}{4}$ inches wide.

Barrels are classified according to the use for which they are fitted, as follows :—

The 1st class includes all barrels, which, if properly coopered, would be fit to carry refined petroleum or naphtha.

The 2nd class includes barrels which are unfit for refined petroleum or naphtha, but which would, if properly coopered, be fit for crude petroleum.

The 3rd class includes such barrels as are unfit for either crude, refined petroleum, or naphtha, but which can be used for residuum, if properly coopered.

179. A tank car weighs complete about 24,500 lbs. Its capacity is 40,000 lbs., the cylinder weigh 6800 lbs. Tank cars have an average capacity for about 100 barrels.

180. Salt water :—

1 cubic foot = 0286 ton = 64·05 lbs. = 1024·84 avd. oz. = 6·2321 gallons.

1 cubic inch = ·0371 lb. = ·5390 avd. oz. = ·0036 gallons.

1 gallon = ·0046 ton = 10·276 lbs. = 164·41 avd. oz. ·1315 cubic feet.

1 ton = 34·973 cubic feet = 2240 lbs. = 217·95 gallons.

NOTE.—A cubic foot of salt water is usually taken at 35 cubic feet to the ton, and 64 lbs. to the cubic foot; fresh water being taken at 36 cubic feet to the ton, and 62·25 lbs. to the cubic foot.—(MACKROW.)

181. Fresh water, specific gravity 1·000.

1 cubic foot = ·0279 ton = 62·39 lbs. = 998·18 avd. oz. = 6·2321 gallons.

1 cubic inch = ·0361 lb. = ·5776 avd. oz. = ·0336 gallon.

1 gallon = ·0045 ton = 10·000 lbs. = 160·15 avd. oz. = ·1315 cubic foot.

1 ton = 35·905 cubic feet = 2240 lbs. = 223·76 gallons.

Weight of fresh water = weight of salt water × ·9740.

—(MACKROW).

1 cubic foot of fresh water, specific gravity 1·000, weighs 62·39 lbs. = 6·23 gallons.

1 gallon of fresh water, specific gravity 1·000, weighs 10·0 lbs.

1 ton of fresh water, specific gravity 1·000, = 35·905 cubic feet = 223·76 gallons.

1 cubic foot sea water, specific gravity 1·027, weighs 64·05 lbs. = 6·23 gallons.

1 gallon sea water, specific gravity 1·027, weighs 10·276 lbs.

1 ton sea water, specific gravity 1·027, = 34·973 cubic feet = 217·95 gallons.

182. Cubic feet = French stères × 35·3156.
French stères = cubic feet × ·02832.
Cubic feet = French litres × ·0353156.
French litres = cubic feet × 28·3161.
Cubic feet = gallons × ·160459.
Gallons = cubic feet × 6·23210.

R

Avoir. lbs. = French kilograms × 2·20462.
French kilograms = avoir. lbs. × ·453593.
Avoir. lbs. = German pfunds × 1·0311.
German pfunds = avoir. lbs. × ·96984.
Tons = Austrian pfunds × ·0005514.
Austrian pfunds = tons × 1813·47.
Tons = French kilograms × ·0009842.
French kilograms = tons × 1016·05.
Poods = tons × 62·03267.
Tons = Russian poods × ·01612453.
1 Russian pood = 36·1141 lbs. avoir.

—(MACKROW.)

1 avd. lb. = ·45359 kilogram.
1 foot = ·304797 metre.
1 square foot = ·092901 square metre.
1 square inch = 645·148 square millimetres.
1 cubic foot = ·028316 cubic metre.
1 cubic yard = ·764534 cubic metre.
1 mile = 1·60933 kilometre.
Knot per hour = 1·688 foot per second.
　　,,　　　,,　　　,, = ·5144 metre per second.
Mile per hour = 1·467 foot per second.
1 gallon = 4·54102 litres.

—(MACKROW.)

183.	Specific gravity	North Atlantic Ocean	..	1·02829
	,,	South ,, ,,	.. ..	1·02882
	,,	Baltic Sea	..	1·01523
	,,	Mediterranean	..	1·02930
	,,	Black Sea	..	1·01418
	,,	Sea of Marmora	..	1·01915
	,,	Red Sea	..	1·0286
	,,	Indian Ocean	..	1·0263

184. Readings on Beaumé's hydrometer compared with ordinary notation.

Beaumé.	Natural Specific Gravity.	Beaumé.	Natural Specific Gravity.	Beaumé.	Natural Specific Gravity.
10°	1·0000	37°	·8383	64°	·7216
11	·9929	38	·8333	65	·7179
12	·9859	39	·8284	66	·7142
13	·9790	40	·8235	67	·7106
14	·9722	41	·8187	68	·7070
15	·9655	42	·8139	69	·7035
16	·9589	43	·8092	70	·7000
17	·9523	44	·8045	71	·6965
18	·9459	45	·8000	72	·6930
19	·9395	46	·7954	73	·6896
20	·9333	47	·7909	74	·6863
21	·9271	48	·7865	75	·6829
22	·9210	49	·7821	76	·6796
23	·9150	50	·7777	77	·6763
24	·9090	51	·7734	78	·6730
25	·9032	52	·7692	79	·6698
26	·8974	53	·7650	80	·6666
27	·8917	54	·7608	81	·6635
28	·8860	55	·7561	82	·6604
29	·8805	56	·7526	83	·6573
30	·8750	57	·7486	84	·6542
31	·8695	58	·7446	85	·6511
32	·8641	59	·7407	86	·6481
33	·8588	60	·7368	87	·6451
34	·8536	61	·7329	88	·6422
35	·8484	62	·7290	89	·6392
36	·8433	63	·7253	90	·6363

185. Thermometer.—Comparisons between the scales of Fahrenheit, Réaumur, and the Centigrade.

Zero Fahrenheit corresponds with minus 17·78 Centigrade and minus 14·22 Réaumur.

Cent.	Fahr.	Rmr.	Cent.	Fahr.	Rmr.	Cent.	Fahr.	Rmr.
°	°	°	°	°	°	°	°	°
100	212	80	75	167	60	50	122	40
99	210·2	79·2	74	165·2	59·2	49	120·2	39·2
98	208·4	78·4	73	163·4	58·4	48	118·4	38·4
97	206·6	77·6	72	161·6	57·6	47	116·6	37·6
96	204·8	76·8	71	159·8	56·8	46	114·8	36·8
95	203	76	70	158	56	45	113	36
94	201·2	75·2	69	156·2	55·2	44	111·2	35·2
93	199·4	74·4	68	154·4	54·4	43	109·4	34·4
92	197·6	73·6	67	152·6	53·6	42	107·6	33·6
91	195·8	72·8	66	150·8	52·8	41	105·8	32·8
90	194	72	65	149	52	40	104	32
89	192·2	71·2	64	147·2	51·2	39	102·2	31·2
88	190·4	70·4	63	145·4	50·4	38	100·4	30·4
87	188·6	69·6	62	143·6	49·6	37	98·6	29·6
86	186·8	68·8	61	141·8	48·8	36	96·8	28·8
85	185	68	60	140	48	35	95	28
84	183·2	67·2	59	138·2	47·2	34	93·2	27·2
83	181·4	66·4	58	136·4	46·4	33	91·4	26·4
82	179·6	65·6	57	134·6	45·6	32	89·6	25·6
81	177·8	64·8	56	132·8	44·8	31	87·8	24·8
80	176	64	55	131	44	30	86	24
79	174·2	63·2	54	129·2	43·2	29	84·2	23·2
78	172·4	62·4	53	127·4	42·4	28	82·4	22·4
77	170·6	61·6	52	125·6	41·6	27	80·6	21·6
76	168·8	60·8	51	123·8	40·8	26	78·8	20·8

THERMOMETER.—COMPARISONS, &c. (*continued*).

Zero Fahrenheit corresponds with minus 17·78 Centigrade and minus 14·22 Réaumur.

Cent.	Fahr.	Rmr.	Cent.	Fahr.	Rmr.	Cent.	Fahr.	Rmr.
°	°	°	°	°	°	°	°	°
25	77	20	Zero	32	Zero	25	13	20
24	75·2	19·2	1	30·2	0·8	26	14·8	20·8
23	73·4	18·4	2	28·4	1·6	27	16·6	21·6
22	71·6	17·6	3	26·6	2·4	28	18·4	22·4
21	69·8	16·8	4	24·8	3·2	29	20·2	23·2
20	68	16	5	23	4	30	22	24
19	66·2	15·2	6	21·2	4·8	31	23·8	24·8
18	64·4	14·4	7	19·4	5·6	32	25·6	25·6
17	62·6	13·6	8	17·6	6·4	33	27·4	26·4
16	60·8	12·8	9	15·8	7 2	34	29·2	27·2
15	59	12	10	14	8	35	31	28
14	57·2	11·2	11	12·2	8·8	36	32·8	28·8
13	55·4	10·4	12	10·4	9·6	37	34·6	29·6
12	53·6	9·6	13	8·6	10·4	38	36·4	30·4
11	51·8	8·8	14	6·8	11·2	39	38·2	31·2
10	50	8	15	5	12	40	40	32
9	48·2	7·2	16	3·2	12·8	41	41·8	32·8
8	46·4	6·4	17	1·4	13·6	42	43·6	33·6
7	44·6	5·6	18	—	14·4	43	45·4	34·4
6	42·8	4·8	19	2·2	15·2	44	47·2	35·2
5	41	4	20	4	16	45	49	36
4	39·2	3·2	21	5·8	16·8	46	50·8	36·8
3	37·4	2·4	22	7·6	17·6	47	52·6	37·6
2	35·6	1·6	23	9·4	18·4	48	54·4	38·4
1	33·8	0·8	24	11·2	19·2	49	56·2	39·2

VOCABULARY OF TECHNICAL TERMS

ENGLISH, GERMAN, AND RUSSIAN.

ENGLISH.	GERMAN.
Yes	Ja.
No	Nein.
All right	Alles wohl !
Go ahead..	Voran ! Vorwärts !
Stop	Halt ! Stop !
Starboard (S. the helm) ..	Steuerbord (Steuerbord das Ruder).
Port (Port the helm)	Backbord (Ruder Backbord).
Valve	Das Ventil, die Klappe.
Pipe .. .. ·	Die Röhre, Leiter, Conductoren.
Nut	Die Mutter, Mutterschraube.
Bolt	Der Bolzen, Klinkbolzen.
Spanner or wrench	Der Schraubendrcher, or Schraubenzicher.
Pump	Die Pumpe.
Engine	Die Maschine, Dampfmaschine.
Steam	Der Dampf.
Tank	Der Behälter, or Wasserkasten.
Petroleum	Das Erdöl, das Steinöl.
Rope or hawser	Das Tau, Kabeltau, Reep, die Pferdclinie.
Tackle	Die Takel, Talje.
Refined petroleum	Refinirtes Erdöl.
Barrel or cask..	Die Tonne, das Fasz.
Case..	Der Koker, Köcher (also Scheide).

VOCABULARY—*continued.*

ENGLISH.	GERMAN.
Flange (of pipe)	Der Flansch, Röhrflansch, *and* das Flanschenröhr = Flange-pipe.
Slowly	Langsam !
Fast	Schnell !
Much slower	Langsamer !
Much faster	Geschwind !
Not so fast	Nicht so schnell.
Not so slow	Nicht so langsam.
Open the valve	Oeffne das Ventil (*or* die Klappe).
Are your valves open ?	Sind die Ventile geöffnet ? Sind die Klappen geöffnet ?
Shut the valve..	Schliesz das Ventil zu (*or* die Klappe).
Are your valves shut ?	Sind die Ventile zugeschlossen ? Sind die Klappen zugeschlossen ?
Commence pumping	Fange-an zu pumpen.
Stop pumping..	Lasz-ab zu pumpen.
Are your pipes connected ? ..	Sind die Röhren verknüpft ?
Disconnect your pipes ..	Knüpfe die Röhren los.
Why do you not pump ? ..	Warum pumpen Sie nicht ?
Is the tank full ?	Ist der Behälter voll ?
Is the tank empty ?	Ist der Behälter leer ?
The tank is full	Der Behälter (*or* Kasten) ist voll.
The tank is empty	Der Behälter (*or* Kasten) ist leer.
All the tanks are full	Alle Behälter sind voll.
All the tanks are empty ..	Alle Behälter sind leer.
The ship is loaded..	Das Schiff ist geladen.
The ship is quite discharged	Das Schiff ist ganz entgeladen.
Are your ballast tanks full ? ..	Sind die Ballastkasten voll ?
Ship	Das Schiff.
Steam-ship	Das Dampfschiff.

VOCABULARY—*continued.*

ENGLISH.	GERMAN.
Boat	Das Boot, Fahrzeug, Kahn.
Ballast-tank	Der Ballastkasten.
How many barrels can you carry ?	Wie viele Tonnen können Sie nehmen ?
Are you ready to take in cargo ?	Sind Sie fertig einzuladen ?
I am ready to take in cargo ..	Ich bin fertig einzuladen.
To-day	Heute.
To-night	Diese Nacht, Heut Abend.
To-morrow	Morgen.
Yesterday	Gestern.
Are your fires out ?	Die Feuer sind sie abgelöscht ?
The fires are not out	Die Feuer sind nicht abgelöscht.
Lamp	Die Lampe.
Electric lamp	Die elektrische Lampe.
Dynamo	Der Dynamo, *or* die Nothglocke.
I sail to-day	Heute gehe ich unter Segel.
I sail to-morrow	Morgen gehe ich unter Segel.
Bill of lading	Das Connossement, Frachtbrief.
Charter-party	Die Charte-partie.
Specific gravity	Spezifische Schwere, *or* Gewicht.
Fire-test	Die Brennprobe.
On demand	Nach Sicht, *or*, bei Vorzeigung.
At sight	A vista.
After sight	Nach Sicht.
After date	Nach Dato, *or*, nach heute.
Pay to the order	Für mich, *or*, uns an die Ordre.
I promise to pay	Werde ich, *or*, werden wir bezahlen.
With interest	Mit Zinsen.

VOCABULARY—*continued.*

ENGLISH.	RUSSIAN.
Yes	Da, tak.
No	Nyet.
All right	Vsyo pravo.
Go ahead	Vpered.
Stop	Stop, ostanovi.
Starboard	Shtirbord.
Port	Bakbord, *or* gavan', *or* pristan.'
Valve	Klapan.
Pipe	Trooba.
Nut	Gaika.
Bolt	Zadvizhka.
Spanner	Otvertka, klyuch.
Pump	Nasos, pompa.
Engine	Mashina, parovaya mashina.
Steam	Par.
Tank	Tender, tsisterna.
Petroleum	Petroleum.
Rope	Verevka.
Tackle	Tali.
Refined petroleum	Kerosin.
Barrel or cask	Barilok, bochka.
Case	Yashchik.
Flange (of pipe)	Greben' trubui.
Slowly	Medlenno.
Fast	Spyeshno.
Much slower	Medlennyei.
Much faster	Spyeshnyei.
Not so fast	Nye tak spyeshno.
Not so slow	Nye tak medlenno.
Open the valve	Otkrui klapan.
Are your valves open?	Otkruitui-li vashi klapanui?
Shut the valve	Zakrui klapan.
Are your valves shut?	Zakruitui-li vashi klapanui?
Commence pumping	Nachinai kachat'.
Stop pumping	Priostanovis' kachat'.
Are your pipes connected?	Obmotanui-li vashi trubui?
Disconnect your pipes	Otmotaite vashi trubui.
Why do you not pump?	Pochemoo Vui ne kachaete?

S

VOCABULARY—*continued.*

ENGLISH.	RUSSIAN.
Is the tank full?	Polon-li tender?
Is the tank empty?	Porozhen-li tender?
The tank is full	Tender uzhe polon.
The tank is empty..	Tender porozhen.
All the tanks are full	Vsye tenderui uzhe polnui.
All the tanks are empty ..	Vsye tenderui porozhenui.
The ship is loaded..	Korabl' yest' nagruzhen.
The ship is quite discharged	Korabl' vpolnye vuigruzhen.
Are your ballast-tanks full?..	Polnui-li vashi balast tenderui?
Ship	Korabl'.
Steamship	Parokhod.
Boat	Lodka.
Ballast-tank	Balast tender.
How many barrels can you carry?	Skalko bochek mozhete prinyat'?
Are you ready to take in cargo?	Gotovui-li vui nagruzhat'?
I am not ready to take in cargo	Ya gotov nagruzhat'.
To-day	Segodnya.
To-night	Segodnya noch'yu.
To-morrow	Zavtra.
Yesterday	Vchera.
Are your fires out?	Pogashenui-li vashi ogni?
The fires are not out	Ogni ne pogashenui.
Lamp	Lampa.
Electric lamp	Elektricheskaya lampa.
Dynamo	Signal'nui kolokol.
I sail to-day	Ya segodnya snimayusya.
I sail to-morrow	Ya zavtra snimayusya.
Bill of lading	Konosament.
Charter-party	Sertepartiya.
Specific gravity	Otnositelnui vyes.
Fire-test	Ognemyer.
On demand	Po bziskam.
At sight	Po prediavieni.
After sight	Po prediavieni.
After date	Gato.
Pay to the order	Nlat it order.
I promise to pay	Ia obetschai.
With interest	Is prozentamu.

VOCABULARY—*continued.*

GERMAN.	ENGLISH.	RUSSIAN.
Ein	1 One	Odun.
Zwei	2 Two	Dba.
Drei	3 Three	Tza.
Vier	4 Four	Tschetire.
Fünf	5 Five	Piat.
Sechs	6 Six	Schest.
Sieben	7 Seven	Sem.
Acht	8 Eight	Votem.
Neun	9 Nine	Deviat.
Zehn	10 Ten	Desat.
Elf	11 Eleven	Odinnatzat.
Zwölf	12 Twelve	Dvenzat.
Dreizehn	13 Thirteen	Trenazat.
Vierzehn	14 Fourteen	Cheterinazat.
Fünfzehn	15 Fifteen ..	Paznatzat.
Sechzehn	16 Sixteen	Schesnadzat.
Siebenzehn	17 Seventeen ..	Semnatzat.
Achtzehn	18 Eighteen	Vosemnatzat.
Neunzehn	19 Nineteen	Davetnazat.
Zwanzig	20 Twenty	Dvatzat.
Ein und Zwanzig ..	21 Twenty-one ..	Dvatzat-odnar.
Dreiszig	30 Thirty	Trudza.
Vierzig	40 Forty	Sorok.
Fünfzig	50 Fifty	Piatdesat.
Sechzig	60 Sixty	Schestdesat.
Siebenzig	70 Seventy	Semdesat.
Achtzig	80 Eighty	Vosemdesat.
Neunzig..	90 Ninety	Devianosto.
Hundert	100 Hundred	Sto.
Tausend	1000 Thousand ..	Tizatz.
Tag	Day	Den.
Woche	Week	Nedala.
Monat	Month	Mesatz.
Jahr	Year	God.

LONDON: PRINTED BY WILLIAM CLOWES AND SONS, LIMITED,
STAMFORD STREET AND CHARING CROSS.

BOOKS RELATING

TO

APPLIED SCIENCE,

PUBLISHED BY

E. & F. N. SPON,

LONDON: 125, STRAND.

NEW YORK: 12, CORTLANDT STREET.

The Engineers' Sketch-Book of Mechanical Move-ments, Devices, Appliances, Contrivances, Details employed in the Design and Construction of Machinery for every purpose. Collected from numerous Sources and from Actual Work. Classified and Arranged for Reference. *Nearly* 2000 *Illustrations.* By T. W. BARBER, Engineer. 8vo, cloth, 7s. 6d.

A Pocket-Book for Chemists, Chemical Manufacturers, Metallurgists, Dyers, Distillers, Brewers, Sugar Refiners, Photographers, Students, etc., etc. By THOMAS BAYLEY, Assoc. R.C. Sc. Ireland, Ana-lytical and Consulting Chemist and Assayer. Fourth edition, with additions, 437 pp., royal 32mo, roan, gilt edges, 5s.

SYNOPSIS OF CONTENTS:

Atomic Weights and Factors—Useful Data—Chemical Calculations—Rules for Indirect Analysis—Weights and Measures—Thermometers and Barometers—Chemical Physics—Boiling Points, etc.—Solubility of Substances—Methods of Obtaining Specific Gravity—Con-version of Hydrometers—Strength of Solutions by Specific Gravity—Analysis—Gas Analysis—Water Analysis—Qualitative Analysis and Reactions—Volumetric Analysis—Manipulation—Mineralogy—Assaying—Alcohol—Beer—Sugar—Miscellaneous Technological matter relating to Potash, Soda, Sulphuric Acid, Chlorine, Tar Products, Petroleum, Milk, Tallow, Photography, Prices, Wages, Appendix, etc., etc.

The Mechanician: A Treatise on the Construction and Manipulation of Tools, for the use and instruction of Young Engineers and Scientific Amateurs, comprising the Arts of Blacksmithing and Forg-ing; the Construction and Manufacture of Hand Tools, and the various Methods of Using and Grinding them; the Construction of Machine Tools, and how to work them; Machine Fitting and Erection; description of Hand and Machine Processes; Turning and Screw Cutting; principles of Constructing and details of Making and Erecting Steam Engines, and the various details of setting out work, etc., etc. By CAMERON KNIGHT, Engineer. *Containing* 1147 *illustrations,* and 397 pages of letter-press, Fourth edition, 4to, cloth, 18s.

B

Just Published, in Demy 8vo, cloth, containing 975 pages and 250 Illustrations, price 7s. 6d.

SPONS' HOUSEHOLD MANUAL:

A Treasury of Domestic Receipts and Guide for Home Management.

PRINCIPAL CONTENTS.

Hints for selecting a good House, pointing out the essential requirements for a good house as to the Site, Soil, Trees, Aspect, Construction, and General Arrangement; with instructions for Reducing Echoes, Waterproofing Damp Walls, Curing Damp Cellars.

Sanitation.—What should constitute a good Sanitary Arrangement; Examples (with illustrations) of Well- and Ill-drained Houses; How to Test Drains; Ventilating Pipes, etc.

Water Supply.—Care of Cisterns; Sources of Supply; Pipes; Pumps; Purification and Filtration of Water.

Ventilation and Warming.—Methods of Ventilating without causing cold draughts, by various means; Principles of Warming; Health Questions; Combustion; Open Grates; Open Stoves; Fuel Economisers; Varieties of Grates; Close-Fire Stoves; Hot-air Furnaces; Gas Heating; Oil Stoves; Steam Heating; Chemical Heaters; Management of Flues; and Cure of Smoky Chimneys.

Lighting.—The best methods of Lighting; Candles, Oil Lamps, Gas, Incandescent Gas, Electric Light; How to test Gas Pipes; Management of Gas.

Furniture and Decoration.—Hints on the Selection of Furniture; on the most approved methods of Modern Decoration; on the best methods of arranging Bells and Calls; How to Construct an Electric Bell.

Thieves and Fire.—Precautions against Thieves and Fire; Methods of Detection; Domestic Fire Escapes; Fireproofing Clothes, etc.

The Larder.—Keeping Food fresh for a limited time; Storing Food without change, such as Fruits, Vegetables, Eggs, Honey, etc.

Curing Foods for lengthened Preservation, as Smoking, Salting, Canning, Potting, Pickling, Bottling Fruits, etc.; Jams, Jellies, Marmalade, etc.

The Dairy.—The Building and Fitting of Dairies in the most approved modern style; Butter-making; Cheesemaking and Curing.

The Cellar.—Building and Fitting; Cleaning Casks and Bottles; Corks and Corking; Aërated Drinks; Syrups for Drinks; Beers; Bitters; Cordials and Liqueurs; Wines; Miscellaneous Drinks.

The Pantry.—Bread-making; Ovens and Pyrometers; Yeast; German Yeast; Biscuits; Cakes; Fancy Breads; Buns.

The Kitchen.—On Fitting Kitchens; a description of the best Cooking Ranges, close and open; the Management and Care of Hot Plates, Baking Ovens, Dampers, Flues, and Chimneys; Cooking by Gas; Cooking by Oil; the Arts of Roasting, Grilling, Boiling, Stewing, Braising, Frying.

Receipts for Dishes—Soups, Fish, Meat, Game, Poultry, Vegetables, Salads, Puddings, Pastry, Confectionery, Ices, etc., etc.; Foreign Dishes.

The Housewife's Room.—Testing Air, Water, and Foods; Cleaning and Renovating; Destroying Vermin.

Housekeeping, Marketing.

The Dining-Room.—Dietetics; Laying and Waiting at Table; Carving; Dinners, Breakfasts, Luncheons, Teas, Suppers, etc.

The Drawing-Room.—Etiquette; Dancing; Amateur Theatricals; Tricks and Illusions; Games (indoor).

The Bedroom and Dressing-Room; Sleep; the Toilet; Dress; Buying Clothes; Outfits; Fancy Dress.

The Nursery.—The Room; Clothing; Washing; Exercise; Sleep; Feeding; Teething; Illness; Home Training.

The Sick-Room.—The Room; the Nurse; the Bed; Sick Room Accessories; Feeding Patients; Invalid Dishes and Drinks; Administering Physic; Domestic Remedies; Accidents and Emergencies; Bandaging; Burns; Carrying Injured Persons; Wounds; Drowning; Fits; Frost-bites; Poisons and Antidotes; Sunstroke; Common Complaints; Disinfection, etc.

The Bath-Room.—Bathing in General; Management of Hot-Water System.

The Laundry.—Small Domestic Washing Machines, and methods of getting up linen; Fitting up and Working a Steam Laundry.

The School-Room.—The Room and its Fittings; Teaching, etc.

The Playground.—Air and Exercise; Training; Outdoor Games and Sports.

The Workroom.—Darning, Patching, and Mending Garments.

The Library.—Care of Books.

The Garden.—Calendar of Operations for Lawn, Flower Garden, and Kitchen Garden.

The Farmyard.—Management of the Horse, Cow, Pig, Poultry, Bees, etc., etc.

Small Motors.—A description of the various small Engines useful for domestic purposes, from 1 man to 1 horse power, worked by various methods, such as Electric Engines, Gas Engines, Petroleum Engines, Steam Engines, Condensing Engines, Water Power, Wind Power, and the various methods of working and managing them.

Household Law.—The Law relating to Landlords and Tenants, Lodgers, Servants, Parochial Authorities, Juries, Insurance, Nuisance, etc.

On Designing Belt Gearing. By E. J. COWLING
WELCH, Mem. Inst. Mech. Engineers, Author of 'Designing Valve Gearing.' Fcap. 8vo, sewed, 6d.

A Handbook of Formulæ, Tables, and Memoranda,
for Architectural Surveyors and others engaged in Building. By J. T. HURST, C.E. Fourteenth edition, royal 32mo, roan, 5s.

" It is no disparagement to the many excellent publications we refer to, to say that in our opinion this little pocket-book of Hurst's is the very best of them all, without any exception. It would be useless to attempt a recapitulation of the contents, for it appears to contain almost *everything* that anyone connected with building could require, and, best of all, made up in a compact form for carrying in the pocket, measuring only 5 in. by 3 in., and about ¾ in. thick, in a limp cover. We congratulate the author on the success of his laborious and practically compiled little book, which has received unqualified and deserved praise from every professional person to whom we have shown it."—*The Dublin Builder.*

Tabulated Weights of Angle, Tee, Bulb, Round,
Square, and Flat Iron and Steel, and other information for the use of Naval Architects and Shipbuilders. By C. H. JORDAN, M.I.N.A. Fourth edition, 32mo, cloth, 2s. 6d.

A Complete Set of Contract Documents for a Country
Lodge, comprising Drawings, Specifications, Dimensions (for quantities), Abstracts, Bill of Quantities, Form of Tender and Contract, with Notes by J. LEANING, printed in facsimile of the original documents, on single sheets fcap., in paper case, 10s.

A Practical Treatise on Heat, as applied to the
Useful Arts; for the Use of Engineers, Architects, &c. By THOMAS BOX. *With 14 plates.* Third edition, crown 8vo, cloth, 12s. 6d.

A Descriptive Treatise on Mathematical Drawing
Instruments: their construction, uses, qualities, selection, preservation, and suggestions for improvements, with hints upon Drawing and Colouring. By W. F. STANLEY, M.R.I. Fifth edition, *with numerous illustrations,* crown 8vo, cloth, 5s.

B **2**

Quantity Surveying. By J. LEANING. With 42 illustrations. Second edition, revised, crown 8vo, cloth, 9s.

CONTENTS :

A complete Explanation of the London Practice.
General Instructions.
Order of Taking Off.
Modes of Measurement of the various Trades.
Use and Waste.
Ventilation and Warming.
Credits, with various Examples of Treatment.
Abbreviations.
Squaring the Dimensions.
Abstracting, with Examples in illustration of each Trade.
Billing.
Examples of Preambles to each Trade.
Form for a Bill of Quantities.
Do.   Bill of Credits.
Do.   Bill for Alternative Estimate.
Restorations and Repairs, and Form of Bill.
Variations before Acceptance of Tender.
Errors in a Builder's Estimate.

Schedule of Prices.
Form of Schedule of Prices.
Analysis of Schedule of Prices.
Adjustment of Accounts.
Form of a Bill of Variations.
Remarks on Specifications.
Prices and Valuation of Work, with Examples and Remarks upon each Trade.
The Law as it affects Quantity Surveyors, with Law Reports.
Taking Off after the Old Method.
Northern Practice.
The General Statement of the Methods recommended by the Manchester Society of Architects for taking Quantities.
Examples of Collections.
Examples of " Taking Off" in each Trade.
Remarks on the Past and Present Methods of Estimating.

Spons' Architects' and Builders' Price Book, with useful Memoranda. Edited by W. YOUNG, Architect. Crown 8vo, cloth, red edges, 3s. 6d. *Published annually.* Sixteenth edition. *Now ready.*

Long-Span Railway Bridges, comprising Investigations of the Comparative Theoretical and Practical Advantages of the various adopted or proposed Type Systems of Construction, with numerous Formulæ and Tables giving the weight of Iron or Steel required in Bridges from 300 feet to the limiting Spans ; to which are added similar Investigations and Tables relating to Short-span Railway Bridges. Second and revised edition. By B. BAKER, Assoc. Inst. C.E. *Plates,* crown 8vo, cloth, 5s.

Elementary Theory and Calculation of Iron Bridges and Roofs. By AUGUST RITTER, Ph.D., Professor at the Polytechnic School at Aix-la-Chapelle. Translated from the third German edition, by H. R. SANKEY, Capt. R.E. With 500 *illustrations*, 8vo, cloth, 15s.

The Elementary Principles of Carpentry. By THOMAS TREDGOLD. Revised from the original edition, and partly re-written, by JOHN THOMAS HURST. Contained in 517 pages of letterpress, and *illustrated with 48 plates and 150 wood engravings.* Sixth edition, reprinted from the third, crown 8vo, cloth, 12s. 6d.

Section I. On the Equality and Distribution of Forces — Section II. Resistance of Timber — Section III. Construction of Floors — Section IV. Construction of Roofs — Section V. Construction of Domes and Cupolas — Section VI. Construction of Partitions — Section VII. Scaffolds, Staging, and Gantries — Section VIII. Construction of Centres for Bridges — Section IX. Coffer-dams, Shoring, and Strutting — Section X. Wooden Bridges and Viaducts — Section XI. Joints, Straps, and other Fastenings — Section XII. Timber.

The Builder's Clerk : a Guide to the Management of a Builder's Business. By THOMAS BALES. Fcap. 8vo, cloth, 1s. 6d.

Practical Gold-Mining: a Comprehensive Treatise on the Origin and Occurrence of Gold-bearing Gravels, Rocks and Ores, and the methods by which the Gold is extracted. By C. G. WARNFORD LOCK, co-Editor of 'Gold, its Occurrence and Extraction.' *With 8 plates and* 271 *engravings in the text,* super-royal 8vo, cloth, 2*l.* 2*s.*

Hot Water Supply: A Practical Treatise upon the Fitting of Circulating Apparatus in connection with Kitchen Range and other Boilers, to supply Hot Water for Domestic and General Purposes. With a Chapter upon Estimating. *Fully illustrated,* crown 8vo, cloth, 3*s.*

Hot Water Apparatus: An Elementary Guide for the Fitting and Fixing of Boilers and Apparatus for the Circulation of Hot Water for Heating and for Domestic Supply, and containing a Chapter upon Boilers and Fittings for Steam Cooking. 32 *illustrations,* fcap. 8vo, cloth, 1*s.* 6*d.*

The Use and Misuse, and the Proper and Improper Fixing of a Cooking Range. Illustrated, fcap. 8vo, sewed, 6*d.*

Iron Roofs: Examples of Design, Description. *Illustrated with* 64 *Working Drawings of Executed Roofs.* By ARTHUR T. WALMISLEY, Assoc. Mem. Inst. C.E. Second edition, revised, imp. 4to, half-morocco, 3*l.* 3*s.*

A History of Electric Telegraphy, to the Year 1837. Chiefly compiled from Original Sources, and hitherto Unpublished Documents, by J. J. FAHIE, Mem. Soc. of Tel. Engineers, and of the International Society of Electricians, Paris. Crown 8vo, cloth, 9*s.*

Spons' Information for Colonial Engineers. Edited by J. T. HURST. Demy 8vo, sewed.

No. 1, Ceylon. By ABRAHAM DEANE, C.E. 2*s.* 6*d.*

CONTENTS:

Introductory Remarks—Natural Productions—Architecture and Engineering—Topography, Trade, and Natural History—Principal Stations—Weights and Measures, etc., etc.

No. 2. Southern Africa, including the Cape Colony, Natal, and the Dutch Republics. By HENRY HALL, F.R.G.S., F.R.C.I. With Map. 3*s.* 6*d.*

CONTENTS:

General Description of South Africa—Physical Geography with reference to Engineering Operations—Notes on Labour and Material in Cape Colony—Geological Notes on Rock Formation in South Africa—Engineering Instruments for Use in South Africa—Principal Public Works in Cape Colony: Railways, Mountain Roads and Passes, Harbour Works, Bridges, Gas Works, Irrigation and Water Supply, Lighthouses, Drainage and Sanitary Engineering, Public Buildings, Mines—Table of Woods in South Africa—Animals used for Draught Purposes—Statistical Notes—Table of Distances—Rates of Carriage, etc.

No. 3. India. By F. C. DANVERS, Assoc. Inst. C.E. With Map. 4*s.* 6*d.*

CONTENTS:

Physical Geography of India—Building Materials—Roads—Railways—Bridges—Irrigation—River Works—Harbours—Lighthouse Buildings—Native Labour—The Principal Trees of India—Money—Weights and Measures—Glossary of Indian Terms, etc.

Our Factories, Workshops, and Warehouses: their Sanitary and Fire-Resisting Arrangements. By B. H. THWAITE, Assoc. Mem. Inst. C.E. *With* 183 *wood engravings,* crown 8vo, cloth, 9*s.*

A Practical Treatise on Coal Mining. By GEORGE G. ANDRÉ, F.G.S., Assoc. Inst. C.E., Member of the Society of Engineers. *With* 82 *lithographic plates.* 2 vols., royal 4to, cloth, 3*l.* 12*s.*

A Practical Treatise on Casting and Founding, including descriptions of the modern machinery employed in the art. By N. E. SPRETSON, Engineer. Third edition, with 82 *plates* drawn to scale, 412 pp., demy 8vo, cloth, 18*s.*

The Depreciation of Factories and their Valuation. By EWING MATHESON, M. Inst. C.E. 8vo, cloth, 6*s.*

A Handbook of Electrical Testing. By H. R. KEMPE, M.S.T.E. Fourth edition, revised and enlarged, crown 8vo, cloth, 16*s.*

Gas Works: their Arrangement, Construction, Plant, and Machinery. By F. COLYER, M. Inst. C.E. *With* 31 *folding plates,* 8vo, cloth, 24*s.*

The Clerk of Works: a Vade-Mecum for all engaged in the Superintendence of Building Operations. By G. G. HOSKINS, F.R.I.B.A. Third edition, fcap. 8vo, cloth, 1*s.* 6*d.*

American Foundry Practice: Treating of Loam, Dry Sand, and Green Sand Moulding, and containing a Practical Treatise upon the Management of Cupolas, and the Melting of Iron. By T. D. WEST, Practical Iron Moulder and Foundry Foreman. Second edition, *with numerous illustrations,* crown 8vo, cloth, 10*s.* 6*d.*

The Maintenance of Macadamised Roads. By T. CODRINGTON, M.I.C.E, F.G.S., General Superintendent of County Roads for South Wales. 8vo, cloth, 6*s.*

Hydraulic Steam and Hand Power Lifting and Pressing Machinery. By FREDERICK COLYER, M. Inst. C.E., M. Inst. M.E. *With* 73 *plates,* 8vo, cloth, 18*s.*

Pumps and Pumping Machinery. By F. COLYER, M.I.C.E., M.I.M.E. *With* 23 *folding plates,* 8vo, cloth, 12*s.* 6*d.*

Pumps and Pumping Machinery. By F. COLYER. Second Part. *With* 11 *large plates,* 8vo, cloth, 12*s.* 6*d.*

A Treatise on the Origin, Progress, Prevention, and Cure of Dry Rot in Timber; with Remarks on the Means of Preserving Wood from Destruction by Sea-Worms, Beetles, Ants, etc. By THOMAS ALLEN BRITTON, late Surveyor to the Metropolitan Board of Works, etc., etc. *With* 10 *plates,* crown 8vo, cloth, 7*s.* 6*d.*

The Municipal and Sanitary Engineer's Handbook.
By H. PERCY BOULNOIS, Mem. Inst. C.E., Borough Engineer, Ports-mouth. *With numerous illustrations*, demy 8vo, cloth, 12s. 6d.

CONTENTS:
The Appointment and Duties of the Town Surveyor—Traffic—Macadamised Roadways—Steam Rolling—Road Metal and Breaking—Pitched Pavements—Asphalte—Wood Pavements—Footpaths—Kerbs and Gutters—Street Naming and Numbering—Street Lighting—Sewer-age—Ventilation of Sewers—Disposal of Sewage—House Drainage—Disinfection—Gas and Water Companies, etc., Breaking up Streets—Improvement of Private Streets—Borrowing Powers—Artizans' and Labourers' Dwellings—Public Conveniences—Scavenging, including Street Cleansing—Watering and the Removing of Snow—Planting Street Trees—Deposit of Plans—Dangerous Buildings—Hoardings—Obstructions—Improving Street Lines—Cellar Openings—Public Pleasure Grounds—Cemeteries—Mortuaries—Cattle and Ordinary Markets—Public Slaughter-houses, etc.—Giving numerous Forms of Notices, Specifications, and General Information upon these and other subjects of great importance to Municipal Engineers and others engaged in Sanitary Work.

Metrical Tables.   By G. L. MOLESWORTH, M.I.C.E.
32mo, cloth, 1s. 6d.

CONTENTS.
General—Linear Measures—Square Measures—Cubic Measures—Measures of Capacity—Weights—Combinations—Thermometers.

Elements of Construction for Electro-Magnets.   By
Count TH. DU MONCEL, Mem. de l'Institut de France. Translated from the French by C. J. WHARTON. Crown 8vo, cloth, 4s. 6d.

Practical Electrical Units Popularly Explained, with
numerous illustrations and Remarks. By JAMES SWINBURNE, late of J. W. Swan and Co., Paris, late of Brush-Swan Electric Light Company, U.S.A. 18mo, cloth, 1s. 6d.

A Treatise on the Use of Belting for the Transmis-sion of Power. By J. H. COOPER. Second edition, *illustrated*, 8vo, cloth, 15s.

A Pocket-Book of Useful Formulæ and Memoranda
for Civil and Mechanical Engineers. By GUILFORD L. MOLESWORTH, Mem. Inst. C.E., Consulting Engineer to the Government of India for State Railways. *With numerous illustrations*, 744 pp. Twenty-second edition, revised and enlarged, 32mo, roan, 6s.

SYNOPSIS OF CONTENTS:
Surveying, Levelling, etc.—Strength and Weight of Materials—Earthwork, Brickwork, Masonry, Arches, etc.—Struts, Columns, Beams, and Trusses—Flooring, Roofing, and Roof Trusses—Girders, Bridges, etc.—Railways and Roads—Hydraulic Formulæ—Canals, Sewers, Waterworks, Docks—Irrigation and Breakwaters—Gas, Ventilation, and Warming—Heat, Light, Colour, and Sound—Gravity: Centres, Forces, and Powers—Millwork, Teeth of Wheels, Shafting, etc.—Workshop Recipes—Sundry Machinery—Animal Power—Steam and the Steam Engine—Water-power, Water-wheels, Turbines, etc.—Wind and Windmills—Steam Navigation, Ship Building, Tonnage, etc.—Gunnery, Projectiles, etc.—Weights, Measures, and Money—Trigonometry, Conic Sections, and Curves—Telegraphy—Mensura-tion—Tables of Areas and Circumference, and Arcs of Circles—Logarithms, Square and Cube Roots, Powers—Reciprocals, etc.—Useful Numbers—Differential and Integral Calcu-lus—Algebraic Signs—Telegraphic Construction and Formulæ.

Hints on Architectural Draughtsmanship. By G. W. TUXFORD HALLATT. Fcap. 8vo, cloth, 1s. 6d.

Spons' Tables and Memoranda for Engineers; selected and arranged by J. T. HURST, C.E., Author of 'Architectural Surveyors' Handbook,' 'Hurst's Tredgold's Carpentry,' etc. Ninth edition, 64mo, roan, gilt edges, 1s.; or in cloth case, 1s. 6d.

This work is printed in a pearl type, and is so small, measuring only 2½ in. by 1¾ in. by ⅜ in. thick, that it may be easily carried in the waistcoat pocket.

"It is certainly an extremely rare thing for a reviewer to be called upon to notice a volume measuring but 2½ in. by 1¾ in., yet these dimensions faithfully represent the size of the handy little book before us. The volume—which contains 118 printed pages, besides a few blank pages for memoranda—is, in fact, a true pocket-book, adapted for being carried in the waistcoat pocket, and containing a far greater amount and variety of information than most people would imagine could be compressed into so small a space. . . . . The little volume has been compiled with considerable care and judgment, and we can cordially recommend it to our readers as a useful little pocket companion."—*Engineering.*

A Practical Treatise on Natural and Artificial Concrete, its Varieties and Constructive Adaptations. By HENRY REID, Author of the 'Science and Art of the Manufacture of Portland Cement.' New Edition, *with 59 woodcuts and 5 plates,* 8vo, cloth, 15s.

Notes on Concrete and Works in Concrete; especially written to assist those engaged upon Public Works. By JOHN NEWMAN, Assoc. Mem. Inst. C.E., crown 8vo, cloth, 4s. 6d.

Electricity as a Motive Power. By Count TH. DU MONCEL, Membre de l Institut de France, and FRANK GERALDY, Ingénieur des Ponts et Chaussées. Translated and Edited, with Additions, by C. J. WHARTON, Assoc. Soc. Tel. Eng. and Elec. *With 113 engravings and diagrams,* crown 8vo, cloth, 7s. 6d.

Treatise on Valve-Gears, with special consideration of the Link-Motions of Locomotive Engines. By Dr. GUSTAV ZEUNER, Professor of Applied Mechanics at the Confederated Polytechnikum of Zurich. Translated from the Fourth German Edition, by Professor J. F. KLEIN, Lehigh University, Bethlehem, Pa. *Illustrated,* 8vo, cloth, 12s. 6d.

The French-Polisher's Manual. By a French-Polisher; containing Timber Staining, Washing, Matching, Improving, Painting, Imitations, Directions for Staining, Sizing, Embodying, Smoothing, Spirit Varnishing, French-Polishing, Directions for Re-polishing. Third edition, royal 32mo, sewed, 6d.

Hops, their Cultivation, Commerce, and Uses in various Countries. By P. L. SIMMONDS. Crown 8vo, cloth, 4s. 6d.

The Principles of Graphic Statics. By GEORGE SYDENHAM CLARKE, Capt. Royal Engineers. *With 112 illustrations.* 4to, cloth, 12s. 6d.

Dynamo-Electric Machinery: A Manual for Students of Electro-technics. By SILVANUS P. THOMPSON, B.A., D.Sc., Professor of Experimental Physics in University College, Bristol, etc., etc. Third edition, *illustrated*, 8vo, cloth, 16s.

Practical Geometry, Perspective, and Engineering Drawing; a Course of Descriptive Geometry adapted to the Require-ments of the Engineering Draughtsman, including the determination of cast shadows and Isometric Projection, each chapter being followed by numerous examples; to which are added rules for Shading, Shade-lining, etc., together with practical instructions as to the Lining, Colouring, Printing, and general treatment of Engineering Drawings, with a chapter on drawing Instruments. By GEORGE S. CLARKE, Capt. R.E. Second edition, *with 21 plates.* 2 vols., cloth, 10s. 6d.

The Elements of Graphic Statics. By Professor KARL VON OTT, translated from the German by G. S. CLARKE, Capt. R.E., Instructor in Mechanical Drawing, Royal Indian Engineering College. *With 93 illustrations,* crown 8vo, cloth, 5s.

A Practical Treatise on the Manufacture and Distri-bution of Coal Gas. By WILLIAM RICHARDS. Demy 4to, with *numerous wood engravings and 29 plates,* cloth, 28s.

SYNOPSIS OF CONTENTS:

Introduction — History of Gas Lighting — Chemistry of Gas Manufacture, by Lewis Thompson, Esq., M.R.C.S.—Coal, with Analyses, by J. Paterson, Lewis Thompson, and G. R. Hislop, Esqrs.—Retorts, Iron and Clay—Retort Setting—Hydraulic Main—Con-densers — Exhausters — Washers and Scrubbers — Purifiers — Purification — History of Gas Holder — Tanks, Brick and Stone, Composite, Concrete, Cast-iron, Compound Annular Wrought-iron — Specifications — Gas Holders — Station Meter — Governor — Distribution—Mains—Gas Mathematics, or Formulæ for the Distribution of Gas, by Lewis Thompson, Esq.—Services—Consumers' Meters—Regulators—Burners—Fittings—Photometer—Carburization of Gas—Air Gas and Water Gas—Composition of Coal Gas, by Lewis Thompson, Esq.—Analyses of Gas—Influence of Atmospheric Pressure and Temperature on Gas—Residual Products—Appendix—Description of Retort Settings, Buildings, etc., etc.

The New Formula for Mean Velocity of Discharge of Rivers and Canals. By W. R. KUTTER. Translated from articles in the 'Cultur-Ingénieur,' by LOWIS D'A. JACKSON, Assoc. Inst. C.E. 8vo, cloth, 12s. 6d.

The Practical Millwright and Engineer's Ready Reckoner; or Tables for finding the diameter and power of cog-wheels, diameter, weight, and power of shafts, diameter and strength of bolts, etc. By THOMAS DIXON. Fourth edition, 12mo, cloth, 3s.

Tin: Describing the Chief Methods of Mining, Dressing and Smelting it abroad; with Notes upon Arsenic, Bismuth and Wolfram. By ARTHUR G. CHARLETON, Mem. American Inst. of Mining Engineers. *With plates,* 8vo. cloth, 12s. 6d.

Perspective, Explained and Illustrated. By G. S. CLARKE, Capt. R.E. *With illustrations,* 8vo, cloth, 3s. 6d.

Practical Hydraulics ; a Series of Rules and Tables for the use of Engineers, etc., etc. By THOMAS BOX. Fifth edition, *numerous plates,* post 8vo, cloth, 5s.

The Essential Elements of Practical Mechanics ; based on the Principle of Work, designed for Engineering Students. By OLIVER BYRNE, formerly Professor of Mathematics, College for Civil Engineers. Third edition, *with* 148 *wood engravings,* post 8vo, cloth, 7s. 6d.

CONTENTS :

Chap. 1. How Work is Measured by a Unit, both with and without reference to a Unit of Time—Chap. 2. The Work of Living Agents, the Influence of Friction, and introduces one of the most beautiful Laws of Motion—Chap. 3. The principles expounded in the first and second chapters are applied to the Motion of Bodies—Chap. 4. The Transmission of Work by simple Machines—Chap. 5. Useful Propositions and Rules.

Breweries and Maltings : their Arrangement, Construction, Machinery, and Plant. By G. SCAMELL, F.R.I.B.A. Second edition, revised, enlarged, and partly rewritten. By F. COLYER, M.I.C.E., M.I.M.E. *With* 20 *plates,* 8vo, cloth, 18s.

A Practical Treatise on the Construction of Horizontal and Vertical Waterwheels, specially designed for the use of operative mechanics. By WILLIAM CULLEN, Millwright and Engineer. *With* 11 *plates.* Second edition, revised and enlarged, small 4to, cloth, 12s. 6d.

A Practical Treatise on Mill-gearing, Wheels, Shafts, Riggers, etc.; for the use of Engineers. By THOMAS BOX. Third edition, *with* 11 *plates.* Crown 8vo, cloth, 7s. 6d.

Mining Machinery: a Descriptive Treatise on the Machinery, Tools, and other Appliances used in Mining. By G. G. ANDRÉ, F.G.S., Assoc. Inst. C.E., Mem. of the Society of Engineers. Royal 4to, uniform with the Author's Treatise on Coal Mining, containing 182 *plates,* accurately drawn to scale, with descriptive text, in 2 vols., cloth, 3l. 12s.

CONTENTS :

Machinery for Prospecting, Excavating, Hauling, and Hoisting—Ventilation—Pumping—Treatment of Mineral Products, including Gold and Silver, Copper, Tin, and Lead, Iron Coal, Sulphur, China Clay, Brick Earth, etc.

Tables for Setting out Curves for Railways, Canals, Roads, etc., varying from a radius of five chains to three miles. By A. KENNEDY and R. W. HACKWOOD. *Illustrated,* 32mo, cloth, 2s. 6d.

The Science and Art of the Manufacture of Portland

Cement, with observations on some of its constructive applications. *With 66 illustrations.* By HENRY REID, C.E., Author of 'A Practical Treatise on Concrete,' etc., etc. 8vo, cloth, 18s.

The Draughtsman's Handbook of Plan and Map

Drawing; including instructions for the preparation of Engineering, Architectural, and Mechanical Drawings. *With numerous illustrations in the text, and 33 plates (15 printed in colours).* By G. G. ANDRÉ, F.G.S., Assoc. Inst. C.E. 4to, cloth, 9s.

CONTENTS:

The Drawing Office and its Furnishings—Geometrical Problems—Lines, Dots, and their Combinations—Colours, Shading, Lettering, Bordering, and North Points—Scales—Plotting—Civil Engineers' and Surveyors' Plans—Map Drawing—Mechanical and Architectural Drawing—Copying and Reducing Trigonometrical Formulæ, etc., etc.

The Boiler-maker's and Iron Ship-builder's Companion,

comprising a series of original and carefully calculated tables, of the utmost utility to persons interested in the iron trades. By JAMES FODEN, author of 'Mechanical Tables,' etc. Second edition revised, *with illustrations,* crown 8vo, cloth, 5s.

Rock Blasting: a Practical Treatise on the means

employed in Blasting Rocks for Industrial Purposes. By G. G. ANDRÉ, F.G.S., Assoc. Inst. C.E. *With 56 illustrations and 12 plates,* 8vo, cloth, 10s. 6d.

Painting and Painters' Manual: a Book of Facts

for Painters and those who Use or Deal in Paint Materials. By C. L. CONDIT and J. SCHELLER. *Illustrated,* 8vo, cloth, 10s. 6d.

A Treatise on Ropemaking as practised in public and

private Rope-yards, with a Description of the Manufacture, Rules, Tables of Weights, etc., adapted to the Trade, Shipping, Mining, Railways, Builders, etc. By R. CHAPMAN, formerly foreman to Messrs. Huddart and Co., Limehouse, and late Master Ropemaker to H.M. Dockyard, Deptford. Second edition, 12mo, cloth, 3s.

Laxton's Builders' and Contractors' Tables; for the

use of Engineers, Architects, Surveyors, Builders, Land Agents, and others. Bricklayer, containing 22 tables, with nearly 30,000 calculations. 4to, cloth, 5s.

Laxton's Builders' and Contractors' Tables. Ex-

cavator, Earth, Land, Water, and Gas, containing 53 tables, with nearly 24,000 calculations. 4to, cloth, 5s.

Egyptian Irrigation. By W. WILLCOCKS, M.I.C.E., Indian Public Works Department, Inspector of Irrigation, Egypt. With Introduction by Lieut.-Col. J. C. ROSS, R.E., Inspector-General of Irrigation. *With numerous lithographs and wood engravings,* royal 8vo, cloth, 1*l.* 16*s.*

Screw Cutting Tables for Engineers and Machinists, giving the values of the different trains of Wheels required to produce Screws of any pitch, calculated by Lord Lindsay, M.P., F.R.S., F.R.A.S., etc. Cloth, oblong, 2*s.*

Screw Cutting Tables, for the use of Mechanical Engineers, showing the proper arrangement of Wheels for cutting the Threads of Screws of any required pitch, with a Table for making the Universal Gas-pipe Threads and Taps. By W. A. MARTIN, Engineer. Second edition, oblong, cloth, 1*s.*, or sewed, 6*d.*

A Treatise on a Practical Method of Designing Slide-Valve Gears by Simple Geometrical Construction, based upon the principles enunciated in Euclid's Elements, and comprising the various forms of Plain Slide-Valve and Expansion Gearing ; together with Stephenson's, Gooch's, and Allan's Link-Motions, as applied either to reversing or to variable expansion combinations. By EDWARD J. COWLING WELCH, Memb. Inst. Mechanical Engineers. Crown 8vo, cloth, 6*s.*

Cleaning and Scouring: a Manual for Dyers, Laundresses, and for Domestic Use. By S. CHRISTOPHER. 18mo, sewed, 6*d.*

A Glossary of Terms used in Coal Mining. By WILLIAM STUKELEY GRESLEY. Assoc. Mem. Inst. C.E., F.G.S., Member of the North of England Institute of Mining Engineers. *Illustrated with numerous woodcuts and diagrams,* crown 8vo, cloth, 5*s.*

A Pocket-Book for Boiler Makers and Steam Users, comprising a variety of useful information for Employer and Workman, Government Inspectors, Board of Trade Surveyors, Engineers in charge of Works and Slips, Foremen of Manufactories, and the general Steam-using Public. By MAURICE JOHN SEXTON. Second edition, royal 32mo, roan, gilt edges, 5*s.*

Electrolysis: a Practical Treatise on Nickeling, Coppering, Gilding, Silvering, the Refining of Metals, and the treatment of Ores by means of Electricity. By HIPPOLYTE FONTAINE, translated from the French by J. A. BERLY, C.E., Assoc. S.T.E. *With engravings.* 8vo, cloth, 9*s.*

Barlow's Tables of Squares, Cubes, Square Roots,

Cube Roots, Reciprocals of all Integer Numbers up to 10,000. Post 8vo, cloth, 6s.

A Practical Treatise on the Steam Engine, con-

taining Plans and Arrangements of Details for Fixed Steam Engines, with Essays on the Principles involved in Design and Construction. By ARTHUR RIGG, Engineer, Member of the Society of Engineers and of the Royal Institution of Great Britain. Demy 4to, *copiously illustrated with woodcuts and* 96 *plates*, in one Volume, half-bound morocco, 2l. 2s.; or cheaper edition, cloth, 25s.

This work is not, in any sense, an elementary treatise, or history of the steam engine, but is intended to describe examples of Fixed Steam Engines without entering into the wide domain of locomotive or marine practice. To this end illustrations will be given of the most recent arrangements of Horizontal, Vertical, Beam, Pumping, Winding, Portable, Semi-portable, Corliss, Allen, Compound, and other similar Engines, by the most eminent Firms in Great Britain and America. The laws relating to the action and precautions to be observed in the construction of the various details, such as Cylinders, Pistons, Piston-rods, Connecting-rods, Cross-heads, Motion-blocks, Eccentrics, Simple, Expansion, Balanced, and Equilibrium Slide-valves, and Valve-gearing will be minutely dealt with. In this connection will be found articles upon the Velocity of Reciprocating Parts and the Mode of Applying the Indicator, Heat and Expansion of Steam Governors, and the like. It is the writer's desire to draw illustrations from every possible source, and give only those rules that present practice deems correct.

A Practical Treatise on the Science of Land and

Engineering Surveying, Levelling, Estimating Quantities, etc., with a general description of the several Instruments required for Surveying, Levelling, Plotting, etc. By H. S. MERRETT. Fourth edition, revised by G. W. USILL, Assoc. Mem. Inst. C.E. 41 *plates, with illustrations and tables*, royal 8vo, cloth, 12s. 6d.

PRINCIPAL CONTENTS :

Part 1. Introduction and the Principles of Geometry. Part 2. Land Surveying; comprising General Observations—The Chain—Offsets Surveying by the Chain only—Surveying Hilly Ground—To Survey an Estate or Parish by the Chain only—Surveying with the Theodolite—Mining and Town Surveying—Railroad Surveying—Mapping—Division and Laying out of Land—Observations on Enclosures—Plane Trigonometry. Part 3. Levelling—Simple and Compound Levelling—The Level Book—Parliamentary Plan and Section—Levelling with a Theodolite—Gradients—Wooden Curves—To Lay out a Railway Curve—Setting out Widths. Part 4. Calculating Quantities generally for Estimates—Cuttings and Embankments—Tunnels—Brickwork—Ironwork—Timber Measuring. Part 5. Description and Use of Instruments in Surveying and Plotting—The Improved Dumpy Level—Troughton's Level—The Prismatic Compass—Proportional Compass—Box Sextant—Vernier—Pantagraph—Merrett's Improved Quadrant—Improved Computation Scale—The Diagonal Scale—Straight Edge and Sector. Part 6. Logarithms of Numbers—Logarithmic Sines and Co-Sines, Tangents and Co-Tangents—Natural Sines and Co-Sines—Tables for Earthwork, for Setting out Curves, and for various Calculations, etc., etc., etc.

Health and Comfort in House Building, or Ventila-

tion with Warm Air by Self-Acting Suction Power, with Review of the mode of Calculating the Draught in Hot-Air Flues, and with some actual Experiments. By J. DRYSDALE, M.D., and J. W. HAYWARD, M.D. Second edition, with Supplement, *with plates*, demy 8vo, cloth, 7s. 6d.

The Assayer's Manual: an Abridged Treatise on the Docimastic Examination of Ores and Furnace and other Artificial Products. By BRUNO KERL. Translated by W. T. BRANNT. *With 65 illustrations,* 8vo, cloth, 12s. 6d.

Dynamo - Electric Machinery: a Text - Book for Students of Electro-Technology. By SILVANUS P. THOMPSON, B.A., D.Sc., M.S.T.E. Third Edition, revised and enlarged, 8vo, cloth, 16s.

The Practice of Hand Turning in Wood, Ivory, Shell, etc., with Instructions for Turning such Work in Metal as may be required in the Practice of Turning in Wood, Ivory, etc.; also an Appendix on Ornamental Turning. (A book for beginners.) By FRANCIS CAMPIN. Third edition, *with wood engravings,* crown 8vo, cloth, 6s.

CONTENTS :

On Lathes—Turning Tools—Turning Wood—Drilling—Screw Cutting—Miscellaneous Apparatus and Processes—Turning Particular Forms—Staining—Polishing—Spinning Metals —Materials—Ornamental Turning, etc.

Treatise on Watchwork, Past and Present. By the Rev. H. L. NELTHROPP, M.A., F.S.A. *With 32 illustrations,* crown 8vo, cloth, 6s. 6d.

CONTENTS :

Definitions of Words and Terms used in Watchwork—Tools—Time—Historical Summary—On Calculations of the Numbers for Wheels and Pinions; their Proportional Sizes, Trains, etc.—Of Dial Wheels, or Motion Work—Length of Time of Going without Winding up—The Verge—The Horizontal—The Duplex—The Lever—The Chronometer—Repeating Watches—Keyless Watches—The Pendulum, or Spiral Spring—Compensation—Jewelling of Pivot Holes—Clerkenwell—Fallacies of the Trade—Incapacity of Workmen—How to Choose and Use a Watch, etc.

Algebra Self-Taught. By W. P. HIGGS, M.A., D.Sc., LL.D., Assoc. Inst. C.E., Author of 'A Handbook of the Differential Calculus,' etc. Second edition, crown 8vo, cloth, 2s. 6d.

CONTENTS :

Symbols and the Signs of Operation—The Equation and the Unknown Quantity— Positive and Negative Quantities—Multiplication—Involution—Exponents—Negative Exponents—Roots, and the Use of Exponents as Logarithms—Logarithms—Tables of Logarithms and Proportionate Parts—Transformation of System of Logarithms—Common Uses of Common Logarithms—Compound Multiplication and the Binomial Theorem—Division, Fractions, and Ratio—Continued Proportion—The Series and the Summation of the Series— Limit of Series—Square and Cube Roots—Equations—List of Formulæ, etc.

Spons' Dictionary of Engineering, Civil, Mechanical, *Military, and Naval;* with technical terms in French, German, Italian, and Spanish, 3100 pp., and *nearly 8000 engravings,* in super-royal 8vo, in 8 divisions, 5l. 8s. Complete in 3 vols., cloth, 5l. 5s. Bound in a superior manner, half-morocco, top edge gilt, 3 vols., 6l. 12s.

Notes in Mechanical Engineering. Compiled principally for the use of the Students attending the Classes on this subject at the City of London College. By HENRY ADAMS, Mem. Inst. M.E., Mem. Inst. C.E., Mem. Soc. of Engineers. Crown 8vo, cloth, 2s. 6d.

Canoe and Boat Building: a complete Manual for Amateurs, containing plain and comprehensive directions for the construction of Canoes, Rowing and Sailing Boats, and Hunting Craft. By W. P. STEPHENS. *With numerous illustrations and 24 plates of Working Drawings.* Crown 8vo, cloth, 7s. 6d.

Proceedings of the National Conference of Electricians, *Philadelphia,* October 8th to 13th, 1884. 18mo, cloth, 3s.

Dynamo - Electricity, its Generation, Application, Transmission, Storage, and Measurement. By G. B. PRESCOTT. *With 545 illustrations.* 8vo, cloth, 1l. 1s.

Domestic Electricity for Amateurs. Translated from the French of E. HOSPITALIER, Editor of "L'Electricien," by C. J. WHARTON, Assoc. Soc. Tel. Eng. *Numerous illustrations.* Demy 8vo, cloth, 9s.

CONTENTS:

1. Production of the Electric Current—2. Electric Bells—3. Automatic Alarms—4. Domestic Telephones—5. Electric Clocks—6. Electric Lighters—7. Domestic Electric Lighting—8. Domestic Application of the Electric Light—9. Electric Motors—10. Electrical Locomotion—11. Electrotyping, Plating, and Gilding—12. Electric Recreations—13. Various applications—Workshop of the Electrician.

Wrinkles in Electric Lighting. By VINCENT STEPHEN. *With illustrations.* 18mo, cloth, 2s. 6d.

CONTENTS:

1. The Electric Current and its production by Chemical means—2. Production of Electric Currents by Mechanical means—3. Dynamo-Electric Machines—4. Electric Lamps—5. Lead—6. Ship Lighting.

The Practical Flax Spinner; being a Description of the Growth, Manipulation, and Spinning of Flax and Tow. By LESLIE C. MARSHALL, of Belfast. *With illustrations.* 8vo, cloth, 15s.

Foundations and Foundation Walls for all classes of *Buildings,* Pile Driving, Building Stones and Bricks, Pier and Wall construction, Mortars, Limes, Cements, Concretes, Stuccos, &c. 64 *illustrations.* By G. T. POWELL and F. BAUMAN. 8vo, cloth, 10s. 6d.

Manual for Gas Engineering Students. By D. LEE.
18mo, cloth 1*s.*

Hydraulic Machinery, Past and Present. A Lecture
delivered to the London and Suburban Railway Officials' Association.
By H. ADAMS, Mem. Inst. C.E. *Folding plate.* 8vo, sewed, 1*s.*

Twenty Years with the Indicator. By THOMAS PRAY,
Jun., C.E., M.E., Member of the American Society of Civil Engineers.
2 vols., royal 8vo, cloth, 12*s.* 6*d.*

Annual Statistical Report of the Secretary to the
Members of the Iron and Steel Association on the Home and Foreign Iron
and Steel Industries in 1887. Issued March 1888. 8vo, sewed, 5*s.*

Bad Drains, and How to Test them; with Notes on
the Ventilation of Sewers, Drains, and Sanitary Fittings, and the Origin
and Transmission of Zymotic Disease. By R. HARRIS REEVES. Crown
8vo, cloth, 3*s.* 6*d.*

Well Sinking. The modern practice of Sinking
and Boring Wells, with geological considerations and examples of Wells.
By ERNEST SPON, Assoc. Mem. Inst. C.E., Mem. Soc. Eng., and of the
Franklin Inst., etc. Second edition, revised and enlarged. Crown 8vo,
cloth, 10*s.* 6*d.*

The Voltaic Accumulator : an Elementary Treatise.
By ÉMILE REYNIER. Translated by J. A. BERLY, Assoc. Inst. E.E.
With 62 *illustrations,* 8vo, cloth, 9*s.*

List of Tests (Reagents), arranged in alphabetical
order, according to the names of the originators. Designed especially
for the convenient reference of Chemists, Pharmacists, and Scientists.
By HANS M. WILDER. Crown 8vo, cloth, 4*s.* 6*d.*

Ten Years' Experience in Works of Intermittent
Downward Filtration. By J. BAILEY DENTON, Mem. Inst. C.E.
Second edition, with additions. Royal 8vo, sewed, 4*s.*

A Treatise on the Manufacture of Soap and Candles,
Lubricants and Glycerin. By W. LANT CARPENTER, B.A., B.Sc. (late
of Messrs. C. Thomas and Brothers, Bristol). *With illustrations.* Crown
8vo, cloth, 10*s.* 6*d.*

The Stability of Ships explained simply, and calculated *by a new Graphic method.* By J. C. SPENCE, M.I.N.A. 4to, sewed, 3s. 6d.

Steam Making, or Boiler Practice. By CHARLES A. SMITH, C.E. 8vo, cloth, 10s. 6d.

CONTENTS:

1. The Nature of Heat and the Properties of Steam—2. Combustion.—3. Externally Fired Stationary Boilers—4. Internally Fired Stationary Boilers—5. Internally Fired Portable Locomotive and Marine Boilers—6. Design, Construction, and Strength of Boilers—7. Proportions of Heating Surface, Economic Evaporation, Explosions—8. Miscellaneous Boilers, Choice of Boiler Fittings and Appurtenances.

The Fireman's Guide; a Handbook on the Care of Boilers. By TEKNOLOG, föreningen T. I. Stockholm. Translated from the third edition, and revised by KARL P. DAHLSTROM, M.E. Second edition. Fcap. 8vo, cloth, 2s.

A Treatise on Modern Steam Engines and Boilers, including Land Locomotive, and Marine Engines and Boilers, for the use of Students. By FREDERICK COLYER, M. Inst. C.E., Mem. Inst. M.E. *With 36 plates.* 4to, cloth, 25s.

CONTENTS:

1. Introduction—2. Original Engines—3. Boilers—4. High-Pressure Beam Engines—5. Cornish Beam Engines—6. Horizontal Engines—7. Oscillating Engines—8. Vertical High-Pressure Engines—9. Special Engines—10. Portable Engines—11. Locomotive Engines—12. Marine Engines.

Steam Engine Management; a Treatise on the Working and Management of Steam Boilers. By F. COLYER, M. Inst. C.E., Mem. Inst. M.E. 18mo, cloth, 2s.

Land Surveying on the Meridian and Perpendicular System. By WILLIAM PENMAN, C.E. 8vo, cloth, 8s. 6d.

The Topographer, his Instruments and Methods, designed for the use of Students, Amateur Photographers, Surveyors, Engineers, and all persons interested in the location and construction of works based upon Topography. *Illustrated with numerous plates, maps, and engravings.* By LEWIS M. HAUPT, A.M. 8vo, cloth, 18s.

A Text-Book of Tanning, embracing the Preparation of all kinds of Leather. By HARRY R. PROCTOR, F.C.S., of Low Lights Tanneries. *With illustrations.* Crown 8vo, cloth, 10s. 6d.

In super-royal 8vo, 1168 pp., *with* 2400 *illustrations*, in 3 Divisions, cloth, price 13*s.* 6*d.* each; or 1 vol., cloth, 2*l.*; or half-morocco, 2*l.* 8*s.*

A SUPPLEMENT

TO

SPONS' DICTIONARY OF ENGINEERING.

EDITED BY ERNEST SPON, MEMB. SOC. ENGINEERS.

Abacus, Counters, Speed Indicators, and Slide Rule.

Agricultural Implements and Machinery.

Air Compressors.

Animal Charcoal Machinery.

Antimony.

Axles and Axle-boxes.

Barn Machinery.

Belts and Belting.

Blasting. Boilers.

Brakes.

Brick Machinery.

Bridges.

Cages for Mines.

Calculus, Differential and Integral.

Canals.

Carpentry.

Cast Iron.

Cement, Concrete, Limes, and Mortar.

Chimney Shafts.

Coal Cleansing and Washing.

Coal Mining.

Coal Cutting Machines.

Coke Ovens. Copper.

Docks. Drainage.

Dredging Machinery.

Dynamo - Electric and Magneto-Electric Machines.

Dynamometers.

Electrical Engineering, Telegraphy, Electric Lighting and its practical details, Telephones

Engines, Varieties of.

Explosives. Fans.

Founding, Moulding and the practical work of the Foundry.

Gas, Manufacture of.

Hammers, Steam and other Power.

Heat. Horse Power.

Hydraulics.

Hydro-geology.

Indicators. Iron.

Lifts, Hoists, and Elevators.

Lighthouses, Buoys, and Beacons.

Machine Tools.

Materials of Construction.

Meters.

Ores, Machinery and Processes employed to Dress.

Piers.

Pile Driving.

Pneumatic Transmission.

Pumps.

Pyrometers.

Road Locomotives.

Rock Drills.

Rolling Stock.

Sanitary Engineering.

Shafting.

Steel.

Steam Navvy.

Stone Machinery.

Tramways.

Well Sinking.

London: E. & F. N. SPON, 125, Strand.

New York: 12, Cortlandt Street.

NOW COMPLETE.

With nearly 1500 *illustrations*, in super-royal 8vo, in 5 Divisions, cloth. Divisions 1 to 4, 13s. 6d. each ; Division 5, 17s. 6d. ; or 2 vols., cloth, £3 10s.

SPONS' ENCYCLOPÆDIA

OF THE

INDUSTRIAL ARTS, MANUFACTURES, AND COMMERCIAL PRODUCTS.

EDITED BY C. G. WARNFORD LOCK, F.L.S.

Among the more important of the subjects treated of, are the following :—

Acids, 207 pp. 220 figs.
Alcohol, 23 pp. 16 figs.
Alcoholic Liquors, 13 pp.
Alkalies, 89 pp. 78 figs.
Alloys.   Alum.
Asphalt.   Assaying.
Beverages, 89 pp. 29 figs.
Blacks.
Bleaching Powder, 15 pp.
Bleaching, 51 pp. 48 figs.
Candles, 18 pp. 9 figs.
Carbon Bisulphide.
Celluloid, 9 pp.
Cements.   Clay.
Coal-tar Products, 44 pp. 14 figs.
Cocoa, 8 pp.
Coffee, 32 pp. 13 figs.
Cork, 8 pp. 17 figs.
Cotton Manufactures, 62 pp. 57 figs.
Drugs, 38 pp.
Dyeing and Calico Printing, 28 pp. 9 figs.
Dyestuffs, 16 pp.
Electro-Metallurgy, 13 pp.
Explosives, 22 pp. 33 figs.
Feathers.
Fibrous Substances, 92 pp. 79 figs.
Floor-cloth, 16 pp. 21 figs.
Food Preservation, 8 pp.
Fruit, 8 pp.

Fur, 5 pp.
Gas, Coal, 8 pp.
Gems.
Glass, 45 pp. 77 figs.
Graphite, 7 pp.
Hair, 7 pp.
Hair Manufactures.
Hats, 26 pp. 26 figs.
Honey.   Hops.
Horn.
Ice, 10 pp. 14 figs.
Indiarubber Manufactures, 23 pp. 17 figs.
Ink, 17 pp.
Ivory.
Jute Manufactures, 11 pp., 11 figs.
Knitted Fabrics — Hosiery, 15 pp. 13 figs.
Lace, 13 pp. 9 figs.
Leather, 28 pp. 31 figs.
Linen Manufactures, 16 pp. 6 figs.
Manures, 21 pp. 30 figs.
Matches, 17 pp. 38 figs.
Mordants, 13 pp.
Narcotics, 47 pp.
Nuts, 10 pp.
Oils and Fatty Substances, 125 pp.
Paint.
Paper, 26 pp. 23 figs.
Paraffin, 8 pp. 6 figs.
Pearl and Coral, 8 pp.
Perfumes, 10 pp.

Photography, 13 pp. 20 figs.
Pigments, 9 pp. 6 figs.
Pottery, 46 pp. 57 figs.
Printing and Engraving, 20 pp. 8 figs.
Rags.
Resinous and Gummy Substances, 75 pp. 16 figs.
Rope, 16 pp. 17 figs.
Salt, 31 pp. 23 figs.
Silk, 8 pp.
Silk Manufactures, 9 pp. 11 figs.
Skins, 5 pp.
Small Wares, 4 pp.
Soap and Glycerine, 39 pp. 45 figs.
Spices, 16 pp.
Sponge, 5 pp.
Starch, 9 pp. 10 figs.
Sugar, 155 pp. 134 figs.
Sulphur.
Tannin, 18 pp.
Tea, 12 pp.
Timber, 13 pp.
Varnish, 15 pp.
Vinegar, 5 pp.
Wax, 5 pp.
Wool, 2 pp.
Woollen Manufactures, 58 pp. 39 figs.

London: E. & F. N. SPON, 125, Strand.
New York: 12, Cortlandt Street.

Crown 8vo, cloth, with illustrations, 5s.

WORKSHOP RECEIPTS,

FIRST SERIES.

BY ERNEST SPON.

SYNOPSIS OF CONTENTS.

Bookbinding.
Bronzes and Bronzing.
Candles.
Cement.
Cleaning.
Colourwashing.
Concretes.
Dipping Acids.
Drawing Office Details.
Drying Oils.
Dynamite.
Electro - Metallurgy — (Cleaning, Dipping, Scratch-brushing, Batteries, Baths, and Deposits of every description).
Enamels.
Engraving on Wood, Copper, Gold, Silver, Steel, and Stone.
Etching and Aqua Tint.
Firework Making — (Rockets, Stars, Rains, Gerbes, Jets, Tourbillons, Candles, Fires, Lances, Lights, Wheels, Fire-balloons, and minor Fireworks).
Fluxes.
Foundry Mixtures.

Freezing.
Fulminates.
Furniture Creams, Oils, Polishes, Lacquers, and Pastes.
Gilding.
Glass Cutting, Cleaning, Frosting, Drilling, Darkening, Bending, Staining, and Painting.
Glass Making.
Glues.
Gold.
Graining.
Gums.
Gun Cotton.
Gunpowder.
Horn Working.
Indiarubber.
Japans, Japanning, and kindred processes.
Lacquers.
Lathing.
Lubricants.
Marble Working.
Matches.
Mortars.
Nitro-Glycerine.
Oils.

Paper.
Paper Hanging.
Painting in Oils, in Water Colours, as well as Fresco, House, Transparency, Sign, and Carriage Painting.
Photography.
Plastering.
Polishes.
Pottery—(Clays, Bodies, Glazes, Colours, Oils, Stains, Fluxes, Enamels, and Lustres).
Scouring.
Silvering.
Soap.
Solders.
Tanning.
Taxidermy.
Tempering Metals.
Treating Horn, Mother-o'-Pearl, and like substances.
Varnishes, Manufacture and Use of.
Veneering.
Washing.
Waterproofing.
Welding.

Besides Receipts relating to the lesser Technological matters and processes, such as the manufacture and use of Stencil Plates, Blacking, Crayons, Paste, Putty, Wax, Size, Alloys, Catgut, Tunbridge Ware, Picture Frame and Architectural Mouldings, Compos, Cameos, and others too numerous to mention.

London: E. & F. N. SPON, 125, Strand.
New York: 12, Cortlandt Street.

Crown 8vo, cloth, 485 pages, with illustrations, 5s.

WORKSHOP RECEIPTS,
SECOND SERIES.

By ROBERT HALDANE.

Pigments, Paint, and Painting: embracing the preparation of *Pigments*, including alumina lakes, blacks (animal, bone, Frankfort, ivory, lamp, sight, soot), blues (antimony, Antwerp, cobalt, cæruleum, Egyptian, manganate, Paris, Péligot, Prussian, smalt, ultramarine), browns (bistre, hinau, sepia, sienna, umber, Vandyke), greens (baryta, Brighton, Brunswick, chrome, cobalt, Douglas, emerald, manganese, mitis, mountain, Prussian, sap, Scheele's, Schweinfurth, titanium, verdigris, zinc), reds (Brazilwood lake, carminated lake, carmine, Cassius purple, cobalt pink, cochineal lake, colcothar, Indian red, madder lake, red chalk, red lead, vermilion), whites (alum, baryta, Chinese, lead sulphate, white lead—by American, Dutch, French, German, Kremnitz, and Pattinson processes, precautions in making, and composition of commercial samples—whiting, Wilkinson's white, zinc white), yellows (chrome, gamboge, Naples, orpiment, realgar, yellow lakes); *Paint* (vehicles, testing oils, driers, grinding, storing, applying, priming, drying, filling, coats, brushes, surface, water-colours, removing smell, discoloration; miscellaneous paints—cement paint for carton-pierre, copper paint, gold paint, iron paint, lime paints, silicated paints, steatite paint, transparent paints, tungsten paints, window paint, zinc paints); *Painting* (general instructions, proportions of ingredients, measuring paint work; carriage painting—priming paint, best putty, finishing colour, cause of cracking, mixing the paints, oils, driers, and colours, varnishing, importance of washing vehicles, re-varnishing, how to dry paint; woodwork painting).

London: E. & F. N. SPON, 125, Strand.
New York: 12, Cortlandt Street.

Crown 8vo, cloth, 480 pages, with 183 illustrations, 5s.

WORKSHOP RECEIPTS,

THIRD SERIES.

By C. G. WARNFORD LOCK.

Uniform with the First and Second Series.

SYNOPSIS OF CONTENTS.

London: E. & F. N. SPON, 125, Strand.
New York: 12, Cortlandt Street.

WORKSHOP RECEIPTS,

FOURTH SERIES,

DEVOTED MAINLY TO HANDICRAFTS & MECHANICAL SUBJECTS.

By C. G. WARNFORD LOCK.

250 Illustrations, with Complete Index, and a General Index to the Four Series, 5s.

Waterproofing — rubber goods, cuprammonium processes, miscellaneous preparations.

Packing and Storing articles of delicate odour or colour, of a deliquescent character, liable to ignition, apt to suffer from insects or damp, or easily broken.

Embalming and Preserving anatomical specimens.

Leather Polishes.

Cooling Air and Water, producing low temperatures, making ice, cooling syrups and solutions, and separating salts from liquors by refrigeration.

Pumps and Siphons, embracing every useful contrivance for raising and supplying water on a moderate scale, and moving corrosive, tenacious, and other liquids.

Desiccating—air- and water-ovens, and other appliances for drying natural and artificial products.

Distilling—water, tinctures, extracts, pharmaceutical preparations, essences, perfumes, and alcoholic liquids.

Emulsifying as required by pharmacists and photographers.

Evaporating—saline and other solutions, and liquids demanding special precautions.

Filtering—water, and solutions of various kinds.

Percolating and Macerating.

Electrotyping.

Stereotyping by both plaster and paper processes.

Bookbinding in all its details.

Straw Plaiting and the fabrication of baskets, matting, etc.

Musical Instruments—the preservation, tuning, and repair of pianos, harmoniums, musical boxes, etc.

Clock and Watch Mending—adapted for intelligent amateurs.

Photography—recent development in rapid processes, handy apparatus, numerous recipes for sensitizing and developing solutions, and applications to modern illustrative purposes.

London : E. & F. N. SPON, 125, Strand.

New York : 12, Cortlandt Street.

In demy 8vo, cloth, 600 pages, and 1420 Illustrations, 6s.

SPONS'

MECHANICS' OWN BOOK;

A MANUAL FOR HANDICRAFTSMEN AND AMATEURS.

CONTENTS.

Mechanical Drawing—Casting and Founding in Iron, Brass, Bronze, and other Alloys—Forging and Finishing Iron—Sheetmetal Working —Soldering, Brazing, and Burning—Carpentry and Joinery, embracing descriptions of some 400 Woods, over 200 Illustrations of Tools and their uses, Explanations (with Diagrams) of 116 joints and hinges, and Details of Construction of Workshop appliances, rough furniture, Garden and Yard Erections, and House Building—Cabinet-Making and Veneering — Carving and Fretcutting — Upholstery — Painting, Graining, and Marbling — Staining Furniture, Woods, Floors, and Fittings—Gilding, dead and bright, on various grounds—Polishing Marble, Metals, and Wood—Varnishing—Mechanical movements, illustrating contrivances for transmitting motion—Turning in Wood and Metals—Masonry, embracing Stonework, Brickwork, Terracotta, and Concrete—Roofing with Thatch, Tiles, Slates, Felt, Zinc, &c.— Glazing with and without putty, and lead glazing—Plastering and Whitewashing — Paper-hanging — Gas-fitting—Bell-hanging, ordinary and electric Systems — Lighting — Warming — Ventilating — Roads, Pavements, and Bridges — Hedges, Ditches, and Drains — Water Supply and Sanitation—Hints on House Construction suited to new countries.

London: E. & F. N. SPON, 125, Strand.
New York: 12, Cortlandt Street.